AN INTRODUCTION TO NONASSOCIATIVE ALGEBRAS

RICHARD D. SCHAFER

Professor of Mathematics, Emeritus
Massachusetts Institute of Technology
Cambridge, Massachusetts

DOVER PUBLICATIONS, INC.
New York

Bibliographical Note

This Dover edition, first published in 1995, is an unabridged, slightly corrected republication of the work first published by Academic Press, Inc., New York, 1966.

Library of Congress Cataloging-in-Publication Data

Schafer, Richard D. (Richard Donald), 1918–
    An introduction to nonassociative algebras / Richard D. Schafer.
      p.   cm.
    Originally published: New York : Academic Press, 1966, in series: Pure and applied mathematics ; 22.
    Includes bibliographical references (p.   –   ) and index.
    ISBN 0-486-68813-5 (pbk.)
    1. Nonassociative algebras. I. Title.
QA252.S33   1995
512′.24—dc20   95-22734
                                                 CIP

Manufactured in the United States of America
Dover Publications, Inc., 31 East 2nd Street, Mineola, N.Y. 11501

To Alice

PREFACE

This little book is an expanded version of the lectures on nonassociative algebras which I gave at an Advanced Subject Matter Institute in Algebra, which was held at Oklahoma State University in the summer of 1961 under the sponsorship of the National Science Foundation.

I have had no desire to write a treatise on this subject. Instead I have tried to present here in an elementary way some topics which have been of interest to me, and which will be helpful to graduate students who are encountering nonassociative algebras for the first time. Proofs are not given of all of the results cited, but a number of the proofs which are included illustrate techniques which are important for the study of non-associative algebras.

Alternative algebras are presented in some detail. I have treated Jordan algebras in a somewhat more cursory way, except for describing their relationships to the exceptional simple Lie algebras. A considerably deeper account of Jordan algebras will be found in the forthcoming book by Jacobson.

I expect that any reader will be acquainted with the content of a beginning course in abstract algebra and linear algebra. Portions of six somewhat more advanced books are recommended for background reading, and at appropriate places reference is made to these books for results concerning quadratic forms, fields, associative algebras, and Lie algebras. The books are:

Albert, A. A., "Structure of Algebras," Vol. 24. American Mathematical Society Colloquium Publications, New York, 1939;

Artin, Emil, " Galois Theory," No. 2, 2nd ed. Notre Dame Mathematical Lectures, Notre Dame, 1948;

Artin, Emil, "Geometric Algebra," No. 3 (Interscience Tracts in Pure and Applied Mathematics). Wiley (Interscience), London and New York, 1957;

Jacobson, Nathan, "Lectures in Abstract Algebra," Vol. II (Linear Algebra). Van Nostrand, Princeton, New Jersey, 1953;

Jacobson, Nathan, "Lie Algebras," No. 10 (Interscience Tracts in Pure and Applied Mathematics). Wiley (Interscience), London and New York, 1962;

Zariski, Oscar, and Samuel, Pierre, "Commutative Algebra," Vol. I. Van Nostrand, Princeton, New Jersey, 1958.

References are also given to some of the research papers listed in the bibliography at the end. It is my hope that this book will serve to make more of the papers cited there accessible to the interested reader.

Completion of this manuscript was partially supported by National Science Foundation Grant GP 2496. I am grateful for this support, and happy to acknowledge it.

RICHARD D. SCHAFER

September, 1966

CONTENTS

I

INTRODUCTION

By common consent a ring $\mathfrak{R}$ is understood to be an additive abelian group in which a multiplication is defined, satisfying

(1.1) $(xy)z = x(yz)$ for all x, y, z in $\mathfrak{R}$

and

(1.2) $(x + y)z = xz + yz, \quad z(x + y) = zx + zy$ for all x, y, z in $\mathfrak{R}$,

while an algebra $\mathfrak{A}$ over a field F is a ring which is a vector space over F with

(1.3) $\alpha(xy) = (\alpha x)y = x(\alpha y)$ for all α in F, x, y in $\mathfrak{A}$,

so that the multiplication in $\mathfrak{A}$ is bilinear. Throughout this book, however, the associative law (1.1) will fail to hold in many of the algebraic systems encountered. For this reason we shall use the terms "ring" and "algebra" for more general systems than customary.

We define a *ring* $\mathfrak{R}$ to be an additive abelian group with a second law of composition, multiplication, which satisfies the distributive laws (1.2). We define an *algebra* $\mathfrak{A}$ over a field F to be a vector space over F with a bilinear multiplication, that is, a multiplication satisfying (1.2) and (1.3). We shall use the name *associative ring* (or *associative algebra*) for a ring (or algebra) in which the associative law (1.1) holds.

In the general literature, an algebra (in our sense) is commonly referred to as a *nonassociative algebra* in order to emphasize that (1.1) is not being assumed. Use of this term does not carry the connotation

1

that (1.1) fails to hold, but only that (1.1) is not assumed to hold. If (1.1) is actually not satisfied in an algebra (or ring), we say that the algebra (or ring) is *not associative*, rather than nonassociative.

As we shall see in Chapter II, a number of basic concepts which are familiar from the study of associative algebras do not involve associativity in any way, and so may fruitfully be employed in the study of nonassociative algebras. For example, we say that two algebras $\mathfrak{A}$ and $\mathfrak{A}'$ over F are *isomorphic* in case there is a vector space isomorphism $x \leftrightarrow x'$ between them with $(xy)' = x'y'$ for all x, y in $\mathfrak{A}$.

Although we shall prove some theorems concerning rings and infinite-dimensional algebras, we shall for the most part be concerned with finite-dimensional algebras. If $\mathfrak{A}$ is an algebra of dimension n over F, let $u_1, \ldots, u_n$ be a basis for $\mathfrak{A}$ over F. Then the bilinear multiplication in $\mathfrak{A}$ is completely determined by the n^3 *multiplication constants* γ_{ijk} which appear in the products

$$(1.4) \qquad u_i u_j = \sum_{k=1}^{n} \gamma_{ijk} u_k, \qquad \gamma_{ijk} \text{ in } F.$$

We shall call the n^2 equations (1.4) a *multiplication table*, and shall sometimes have occasion to arrange them in the familiar form of such a table:

$$
\begin{array}{c|ccccc}
 & u_1 & \cdots & u_j & \cdots & u_n \\
\hline
u_1 & & & \vdots & & \\
\vdots & & & \vdots & & \\
u_i & \cdots\cdots & \sum \gamma_{ijk} u_k & \cdots\cdots \\
\vdots & & & \vdots & & \\
u_n & & & \vdots & & \\
\end{array}
$$

The multiplication table for a one-dimensional algebra $\mathfrak{A}$ over F is given by $u_1^2 = \gamma u_1$ ($\gamma = \gamma_{111}$). There are two cases: $\gamma = 0$ (from which it follows that every product xy in $\mathfrak{A}$ is 0, so that $\mathfrak{A}$ is called a *zero algebra*), and $\gamma \neq 0$. In the latter case the element $e = \gamma^{-1} u_1$ serves as a basis for $\mathfrak{A}$ over F, and in the new multiplication table we have $e^2 = e$. Then $\alpha \leftrightarrow \alpha e$ is an isomorphism between F and this one-dimensional algebra $\mathfrak{A}$. We have seen incidentally that any one-dimensional algebra is associative. There is much more variety, however, among the algebras which can be encountered even for such a low dimension as two.

Other than associative algebras, the best-known examples of algebras are the Lie algebras which arise in the study of Lie groups. A *Lie algebra* $\mathfrak{L}$ over F is an algebra over F in which the multiplication is *anticommutative*, that is,

$$x^2 = 0 \qquad \text{(implying } xy = -yx\text{)},$$

and the *Jacobi identity*

$$(xy)z + (yz)x + (zx)y = 0 \qquad \text{for all} \quad x, y, z \text{ in } \mathfrak{L}$$

is satisfied. If $\mathfrak{A}$ is any associative algebra over F, then the *commutator*

(1.5) $$[x, y] = xy - yx$$

satisfies

$$[x, x] = 0 \qquad \text{and} \qquad [[x, y], z] + [[y, z], x] + [[z, x], y] = 0.$$

Thus the algebra $\mathfrak{A}^-$ obtained by defining a new multiplication (1.5) in the same vector space as $\mathfrak{A}$ is a Lie algebra over F. Also any subspace of $\mathfrak{A}$ which is closed under commutation (1.5) gives a subalgebra of $\mathfrak{A}^-$, hence a Lie algebra over F. For example, if $\mathfrak{A}$ is the associative algebra of all $n \times n$ matrices, then the set $\mathfrak{L}$ of all skew-symmetric matrices in $\mathfrak{A}$ is a Lie algebra of dimension $\frac{1}{2}n(n - 1)$. The Birkhoff–Witt theorem states that any Lie algebra $\mathfrak{L}$ is isomorphic to a subalgebra of an (infinite-dimensional) algebra $\mathfrak{A}^-$ where $\mathfrak{A}$ is associative (Jacobson [25], pp. 159–162). In the general literature the notation $[x, y]$, without regard to (1.5), is frequently used, instead of xy, to denote the product in an arbitrary Lie algebra.

In this book we shall not make any systematic study of Lie algebras. A number of such accounts exist (principally for characteristic 0, where most of the known results lie); we shall refer, in particular, as above to Jacobson [25]. Instead we shall be concerned upon occasion with relationships between Lie algebras and other nonassociative algebras which arise through such mechanisms as the *derivation algebra*. Let $\mathfrak{A}$ be any algebra over F. By a *derivation* of $\mathfrak{A}$ is meant a linear operator D on $\mathfrak{A}$ satisfying

$$(xy)D = (xD)y + x(yD) \qquad \text{for all} \quad x, y \text{ in } \mathfrak{A}.$$

The set $\mathfrak{D}(\mathfrak{A})$ of all derivations of $\mathfrak{A}$ is a subspace of the associative algebra $\mathfrak{E} = \mathfrak{E}(\mathfrak{A})$ of all linear operators on $\mathfrak{A}$. Since the commutator $[D, D']$ of two derivations D, D' is a derivation of $\mathfrak{A}$, $\mathfrak{D}(\mathfrak{A})$ is a subalgebra of $\mathfrak{E}^-$; that is, $\mathfrak{D}(\mathfrak{A})$ is a Lie algebra, called the *derivation algebra* of $\mathfrak{A}$.

Just as one can introduce the commutator (1.5) as a new product to obtain a Lie algebra $\mathfrak{A}^-$ from an associative algebra $\mathfrak{A}$, so one can introduce a symmetrized product

(1.6) $x * y = xy + yx$

in an associative algebra $\mathfrak{A}$ to obtain a new algebra over F where the vector space operations coincide with those in $\mathfrak{A}$ but where multiplication is defined by the commutative product $x * y$. If one is content to restrict attention to fields F of characteristic not two (as we shall be in many places), there is a certain advantage in writing

(1.7) $x \cdot y = \frac{1}{2}(xy + yx)$

to obtain an algebra $\mathfrak{A}^+$ from an associative algebra $\mathfrak{A}$ by defining products by (1.7) in the same vector space as $\mathfrak{A}$. For $\mathfrak{A}^+$ is isomorphic under the mapping $a \to \frac{1}{2}a$ to the algebra in which products are defined by (1.6). At the same time, powers of any element x in $\mathfrak{A}^+$ coincide with those in $\mathfrak{A}$: clearly $x \cdot x = x^2$, whence it is easy to see by induction on n that

$$x \cdot x \cdot \cdots \cdot x \ (n \text{ factors}) = (x \cdot \cdots \cdot x) \cdot (x \cdot \cdots \cdot x)$$
$$= x^i \cdot x^{n-i} = \tfrac{1}{2}(x^i x^{n-i} + x^{n-i} x^i) = x^n.$$

If $\mathfrak{A}$ is associative, then the multiplication in $\mathfrak{A}^+$ is not only commutative but also satisfies the identity

$$(x \cdot y) \cdot (x \cdot x) = x \cdot [y \cdot (x \cdot x)] \qquad \text{for all} \quad x, y \text{ in } \mathfrak{A}^+.$$

A (commutative) *Jordan algebra* $\mathfrak{J}$ is an algebra over a field F in which products are *commutative*:

$$xy = yx \qquad \text{for all} \quad x, y \text{ in } \mathfrak{J},$$

and satisfy the *Jordan identity*

$$(xy)x^2 = x(yx^2) \qquad \text{for all} \quad x, y \text{ in } \mathfrak{J}.$$

Thus, if $\mathfrak{A}$ is associative, then $\mathfrak{A}^+$ is a Jordan algebra. So is any sub-algebra of $\mathfrak{A}^+$, that is, any subspace of $\mathfrak{A}$ which is closed under the symmetrized product (1.7) and in which (1.7) is used as a new multiplication (for example, the set of all $n \times n$ symmetric matrices). An algebra $\mathfrak{J}$ over F is called a *special Jordan algebra* in case $\mathfrak{J}$ is isomorphic to a subalgebra of $\mathfrak{A}^+$ for some associative $\mathfrak{A}$. We shall see that not all Jordan algebras are special.

Jordan algebras were introduced in the early 1930's by a physicist, P. Jordan, in an attempt to generalize the formalism of quantum mechanics. Little appears to have resulted in this direction, but un-anticipated relationships between these algebras and Lie groups and the foundations of geometry have been discovered.

The study of Jordan algebras which are not special depends upon knowledge of a class of algebras which are more general, but in a certain sense only slightly more general, than associative algebras. These are the *alternative* algebras $\mathfrak{A}$ defined by the identities

$$x^2 y = x(xy) \qquad \text{for all} \quad x, y \text{ in } \mathfrak{A}$$

and

$$yx^2 = (yx)x \qquad \text{for all} \quad x, y \text{ in } \mathfrak{A},$$

known respectively as the *left* and *right alternative laws*. Clearly, any associative algebra is alternative. The class of 8-dimensional *Cayley algebras* (or *Cayley–Dickson algebras*, the prototype having been discovered in 1845 by Cayley and later generalized by Dickson) is, as we shall see, an important class of alternative algebras which are not associative.

Let F have characteristic $\neq 2$. Then the multiplication table for any Cayley algebra $\mathfrak{C}$ over F may be taken to be:

	u_1	u_2	u_3	u_4	u_5	u_6	u_7	u_8
u_1	u_1	u_2	u_3	u_4	u_5	u_6	u_7	u_8
u_2	u_2	$\mu_1 u_1$	$-u_4$	$-\mu_1 u_3$	$-u_6$	$-\mu_1 u_5$	u_8	$\mu_1 u_7$
u_3	u_3	u_4	$\mu_2 u_1$	$\mu_2 u_2$	$-u_7$	$-u_8$	$-\mu_2 u_5$	$-\mu_2 u_6$
u_4	u_4	$\mu_1 u_3$	$-\mu_2 u_2$	$-\mu_1\mu_2 u_1$	$-u_8$	$-\mu_1 u_7$	$\mu_2 u_6$	$\mu_1\mu_2 u_5$
u_5	u_5	u_6	u_7	u_8	$\mu_3 u_1$	$\mu_3 u_2$	$\mu_3 u_3$	$\mu_3 u_4$
u_6	u_6	$\mu_1 u_5$	u_8	$\mu_1 u_7$	$-\mu_3 u_2$	$-\mu_1\mu_3 u_1$	$-\mu_3 u_4$	$-\mu_1\mu_3 u_3$
u_7	u_7	$-u_8$	$\mu_2 u_5$	$-\mu_2 u_6$	$-\mu_3 u_3$	$\mu_3 u_4$	$-\mu_2\mu_3 u_1$	$\mu_2\mu_3 u_2$
u_8	u_8	$-\mu_1 u_7$	$\mu_2 u_6$	$-\mu_1\mu_2 u_5$	$-\mu_3 u_4$	$\mu_1\mu_3 u_3$	$-\mu_2\mu_3 u_2$	$\mu_1\mu_2\mu_3 u_1$

where $u_1, u_2, ..., u_8$ is a basis for $\mathfrak{C}$ over F, and μ_1, μ_2, and μ_3 are nonzero elements of F.

To date these are the algebras (Lie, Jordan, and alternative) about which most is known. Numerous generalizations have recently been made, usually by studying classes of algebras defined by weaker identities.

The structure theories for associative and Lie algebras have served as models for generalization and analogy. Let us recapitulate here some well-known features of these theories.

Let F be an arbitrary field and $\mathfrak{A}$ be a finite-dimensional associative algebra over F. As one learns from, for example, Albert [24], there is an ideal $\mathfrak{N}$, called the radical of $\mathfrak{A}$, which is the unique maximal nilideal of $\mathfrak{A}$ (that is, the maximal ideal consisting entirely of nilpotent elements). Furthermore, $\mathfrak{N}$ is nilpotent in the sense that there is an integer t with the property that any product $z_1 z_2 \cdots z_t$ of t elements from $\mathfrak{N}$ is zero; hence $\mathfrak{N}$ is also the unique maximal nilpotent ideal of $\mathfrak{A}$. Modulo this radical the algebra is semisimple; that is, the residue class algebra $\mathfrak{A}/\mathfrak{N}$ has radical equal to zero. Moreover, any semisimple associative algebra is uniquely expressible as a direct sum $\mathfrak{S}_1 \oplus \cdots \oplus \mathfrak{S}_r$ of simple two-sided ideals (where an algebra is simple provided it has no proper ideals and is not a 1-dimensional algebra in which all products are zero). A celebrated theorem of Wedderburn states that any simple associative algebra $\mathfrak{S}$ is the Kronecker product $\mathfrak{D}_n = \mathfrak{D} \otimes F_n$ of a division algebra $\mathfrak{D}$ over F and the total matrix algebra F_n of dimension n^2, where n is unique and $\mathfrak{D}$ is uniquely determined up to isomorphism. Hence (up to a determination of all division algebras $\mathfrak{D}$ over F) the structure of any semisimple associative algebra over F is known.

Let $\mathfrak{A}/\mathfrak{N}$ be separable (that is, the center of each simple component is a separable field over F; this would always be the case if F were of characteristic 0). Then $\mathfrak{A}$ has a Wedderburn decomposition $\mathfrak{A} = \mathfrak{S} + \mathfrak{N}$ where $\mathfrak{S}$ is a subalgebra of $\mathfrak{A}$ isomorphic to $\mathfrak{A}/\mathfrak{N}$ and $\mathfrak{S} + \mathfrak{N}$ is a vector space direct sum. Any derivation D of a separable algebra $\mathfrak{A}$ is inner: there exists x in $\mathfrak{A}$ such that $aD = ax - xa$ for all a in $\mathfrak{A}$ (Jacobson [2]).

This of course has been but the briefest of sketches of the associative structure theory, and omits many important features. It is astonishing, however, how closely the structure of Lie algebras of characteristic 0 parallels the associative theory up to this point (Jacobson [25]).

Let F be a field of characteristic 0 and $\mathfrak{L}$ be a finite-dimensional Lie algebra over F. Then the radical $\mathfrak{N}$ of $\mathfrak{L}$ is not the maximal nilideal (since $\mathfrak{L}$ itself is a nilalgebra, the square of every element being zero), nor is it in general the maximal nilpotent ideal of $\mathfrak{L}$. It is an ideal between these two. Define $\mathfrak{B}^{(1)} = \mathfrak{B}$, $\mathfrak{B}^{(i+1)} = (\mathfrak{B}^{(i)})^2$. Then $\mathfrak{B}$ is solvable in case there is an integer r such that $\mathfrak{B}^{(r)} = 0$, and the radical $\mathfrak{N}$ of $\mathfrak{L}$ is the unique maximal solvable ideal of $\mathfrak{L}$. With this definition of radical, the residue class algebra $\mathfrak{L}/\mathfrak{N}$ is semisimple and is uniquely expressible as a direct sum of simple two-sided ideals. If F is algebraically closed, the classification of simple Lie algebras into four great classes and five exceptional algebras is well known. This leads to a determination of the simple Lie algebras over arbitrary F of characteristic 0 which by now is almost complete, and in this sense we can say that all semisimple Lie algebras over F are known.

Since F is of characteristic 0, there is no question of separability involved. Any finite-dimensional Lie algebra $\mathfrak{L}$ over F has a Levi decomposition $\mathfrak{L} = \mathfrak{S} + \mathfrak{N}$ where $\mathfrak{S}$ is a subalgebra isomorphic to $\mathfrak{L}/\mathfrak{N}$ and $\mathfrak{S} + \mathfrak{N}$ is a vector space direct sum. Any derivation D of a semisimple algebra $\mathfrak{L}$ is inner: $aD = ax$ for some x in $\mathfrak{L}$ ($D = \mathrm{ad}\ x$).

Lest we be too taken up with the similarities between these two theories, we should perhaps recall one of the important tools for the study of associative algebras; namely, the Peirce decomposition relative to an idempotent. Let e be an idempotent ($e^2 = e \neq 0$) in an associative algebra $\mathfrak{A}$ over an arbitrary field F. Then $\mathfrak{A}$ may be written as the vector space direct sum $\mathfrak{A} = \mathfrak{A}_{11} + \mathfrak{A}_{10} + \mathfrak{A}_{01} + \mathfrak{A}_{00}$ of spaces $\mathfrak{A}_{ij}$, which consist of those elements x_{ij} in $\mathfrak{A}$ satisfying $ex_{ij} = ix_{ij}$, $x_{ij}e = jx_{ij}$ ($i, j = 0, 1$). The properties of this decomposition are essential to the proofs of some of the associative theorems mentioned. On the other hand, there are clearly no idempotents in Lie algebras, and the proofs of the parallel theorems rely on other methods, notably on a trace argument which breaks down for fields of characteristic $p > 0$.

We shall prove in Chapter III that each of these results about associative algebras is a particular case of a corresponding generalization to finite-dimensional alternative algebras. In Chapter IV, although not all of the proofs are included, we shall see that the analogues of these theorems about associative and Lie algebras hold also for finite-dimensional Jordan algebras of characteristic $\neq 2$.

For broader classes of nonassociative algebras the associative and Lie theories do not serve as appropriate models; counterexamples to these theorems exist. We shall see in Chapter II, however, some nontrivial results which hold for completely arbitrary algebras.

II

ARBITRARY NONASSOCIATIVE ALGEBRAS

1. SOME BASIC CONCEPTS

Let $\mathfrak{A}$ be an algebra over a field F. (The reader may make the appropriate modifications for a ring $\mathfrak{R}$.) The definitions of the terms *subalgebra, left ideal, right ideal,* (two-sided) *ideal* $\mathfrak{B}$, *homomorphism, kernel* of a homomorphism, *residue class algebra* $\mathfrak{A}/\mathfrak{B}$ (difference algebra $\mathfrak{A}$-$\mathfrak{B}$), *anti-isomorphism*, which are familiar from a study of associative algebras, do not involve associativity of multiplication and are thus immediately applicable to algebras in general. So is the notation $\mathfrak{B}\mathfrak{C}$ for the subspace of $\mathfrak{A}$ spanned by all products bc with b in $\mathfrak{B}$, c in $\mathfrak{C}$ ($\mathfrak{B}$, $\mathfrak{C}$ being arbitrary nonempty subsets of $\mathfrak{A}$); here we must of course distinguish between $(\mathfrak{A}\mathfrak{B})\mathfrak{C}$ and $\mathfrak{A}(\mathfrak{B}\mathfrak{C})$, etc.

We have the *fundamental theorem of homomorphism for algebras*: If $\mathfrak{B}$ is an ideal of $\mathfrak{A}$, then $\mathfrak{A}/\mathfrak{B}$ is a homomorphic image of $\mathfrak{A}$ under the natural homomorphism

$$a \to \bar{a} = a + \mathfrak{B}, \qquad a \text{ in } \mathfrak{A}, \quad a + \mathfrak{B} \text{ in } \mathfrak{A}/\mathfrak{B}.$$

Conversely, if $\mathfrak{A}'$ is a homomorphic image of $\mathfrak{A}$ (under the homomorphism

$$(2.1) \qquad\qquad a \to a', \qquad a \text{ in } \mathfrak{A}, \quad a' \text{ in } \mathfrak{A}'),$$

9

then $\mathfrak{A}'$ is isomorphic to $\mathfrak{A}/\mathfrak{B}$ where $\mathfrak{B}$ is the kernel of the homomorphism.

If $\mathfrak{S}'$ is a subalgebra (or ideal) of a homomorphic image $\mathfrak{A}'$ of $\mathfrak{A}$, then the *complete inverse image* of $\mathfrak{S}'$ under the homomorphism (2.1)—that is, the set $\mathfrak{S} = \{s \in \mathfrak{A} \,|\, s' \in \mathfrak{S}'\}$—is a subalgebra (or ideal) of $\mathfrak{A}$ which contains the kernel $\mathfrak{B}$ of (2.1). If a class of algebras is defined by identities (as, for example, Lie, Jordan or alternative algebras), then any subalgebra or any homomorphic image belongs to the same class.

We have the customary isomorphism theorems:

(i) If $\mathfrak{B}_1$ and $\mathfrak{B}_2$ are ideals of $\mathfrak{A}$ such that $\mathfrak{B}_1$ contains $\mathfrak{B}_2$, then $(\mathfrak{A}/\mathfrak{B}_2)/(\mathfrak{B}_1/\mathfrak{B}_2)$ and $\mathfrak{A}/\mathfrak{B}_1$ are isomorphic.

(ii) If $\mathfrak{B}$ is an ideal of $\mathfrak{A}$ and $\mathfrak{S}$ is a subalgebra of $\mathfrak{A}$, then $\mathfrak{B} \cap \mathfrak{S}$ is an ideal of $\mathfrak{S}$, and $(\mathfrak{B} + \mathfrak{S})/\mathfrak{B}$ and $\mathfrak{S}/(\mathfrak{B} \cap \mathfrak{S})$ are isomorphic.

Suppose that $\mathfrak{B}$ and $\mathfrak{C}$ are ideals of an algebra $\mathfrak{A}$, and that as a vector space $\mathfrak{A}$ is the direct sum of $\mathfrak{B}$ and $\mathfrak{C}$ ($\mathfrak{A} = \mathfrak{B} + \mathfrak{C}$, $\mathfrak{B} \cap \mathfrak{C} = 0$). Then $\mathfrak{A}$ is called the *direct sum* $\mathfrak{A} = \mathfrak{B} \oplus \mathfrak{C}$ of $\mathfrak{B}$ and $\mathfrak{C}$ as algebras. The vector space properties ensure that in a direct sum $\mathfrak{A} = \mathfrak{B} \oplus \mathfrak{C}$ the components b, c of $a = b + c$ (b in $\mathfrak{B}$, c in $\mathfrak{C}$) are uniquely determined, and that addition and multiplication by scalars are performed componentwise. It is the assumption that $\mathfrak{B}$ and $\mathfrak{C}$ are ideals in $\mathfrak{A} = \mathfrak{B} \oplus \mathfrak{C}$ that gives componentwise multiplication as well:

$$(b_1 + c_1)(b_2 + c_2) = b_1 b_2 + c_1 c_2, \qquad b_i \text{ in } \mathfrak{B}, \quad c_i \text{ in } \mathfrak{C}.$$

For $b_1 c_2$ is in both $\mathfrak{B}$ and $\mathfrak{C}$, hence in $\mathfrak{B} \cap \mathfrak{C} = 0$. Similarly $c_1 b_2 = 0$. (Although $\oplus$ is commonly used to denote vector space direct sum, it has been reserved here for direct sum of ideals; where appropriate the notation $\perp$ has been used for orthogonal direct sum relative to a symmetric bilinear form.)

Given any two algebras $\mathfrak{B}$, $\mathfrak{C}$ over a field F, one can construct an algebra $\mathfrak{A}$ over F such that $\mathfrak{A}$ is the direct sum $\mathfrak{A} = \mathfrak{B}' \oplus \mathfrak{C}'$ of ideals $\mathfrak{B}'$, $\mathfrak{C}'$ which are isomorphic respectively to $\mathfrak{B}$, $\mathfrak{C}$. The construction of $\mathfrak{A}$ is familiar: the elements of $\mathfrak{A}$ are the ordered pairs (b, c) with b in $\mathfrak{B}$, c in $\mathfrak{C}$; addition, multiplication by scalars, and multiplication are defined componentwise:

$$(b_1, c_1) + (b_2, c_2) = (b_1 + b_2, c_1 + c_2),$$
$$\alpha(b, c) = (\alpha b, \alpha c),$$
$$(b_1, c_1)(b_2, c_2) = (b_1 b_2, c_1 c_2).$$

Then $\mathfrak{A}$ is an algebra over F, the sets $\mathfrak{B}'$ of all pairs $(b, 0)$ with b in $\mathfrak{B}$ and $\mathfrak{C}'$ of all pairs $(0, c)$ with c in $\mathfrak{C}$ are ideals of $\mathfrak{A}$ isomorphic respectively to $\mathfrak{B}$ and $\mathfrak{C}$, and $\mathfrak{A} = \mathfrak{B}' \oplus \mathfrak{C}'$. By the customary identification of $\mathfrak{B}$ with $\mathfrak{B}'$, $\mathfrak{C}$ with $\mathfrak{C}'$, we can then write $\mathfrak{A} = \mathfrak{B} \oplus \mathfrak{C}$, the direct sum of $\mathfrak{B}$ and $\mathfrak{C}$ as algebras.

As in the case of vector spaces, the notion of direct sum extends to an arbitrary (indexed) set of summands. In this book we shall have occasion to use only finite direct sums $\mathfrak{A} = \mathfrak{B}_1 \oplus \mathfrak{B}_2 \oplus \cdots \oplus \mathfrak{B}_t$. Here $\mathfrak{A}$ is the direct sum of the vector spaces $\mathfrak{B}_i$, and multiplication in $\mathfrak{A}$ is given by

$$(b_1 + b_2 + \cdots + b_t)(c_1 + c_2 + \cdots + c_t) = b_1 c_1 + b_2 c_2 + \cdots + b_t c_t$$

for b_i, c_i in $\mathfrak{B}_i$. The $\mathfrak{B}_i$ are ideals of $\mathfrak{A}$. Note that (in the case of a vector space direct sum) the latter statement is equivalent to the fact that the $\mathfrak{B}_i$ are subalgebras of $\mathfrak{A}$ such that

$$\mathfrak{B}_i \mathfrak{B}_j = 0 \qquad \text{for} \quad i \neq j.$$

An element e (or f) in an algebra $\mathfrak{A}$ over F is called a *left* (or *right*) *identity* (sometimes *unity element*) in case $ea = a$ (or $af = a$) for all a in $\mathfrak{A}$. If $\mathfrak{A}$ contains both a left identity e and a right identity f, then $e = f$ ($= ef$) is a (two-sided) *identity* 1. If $\mathfrak{A}$ does not contain an identity element 1, there is a standard construction for obtaining an algebra $\mathfrak{A}_1$ which does contain 1, such that $\mathfrak{A}_1$ contains (an isomorphic copy of) $\mathfrak{A}$ as an ideal, and such that $\mathfrak{A}_1/\mathfrak{A}$ has dimension 1 over F. We take $\mathfrak{A}_1$ to be the set of all ordered pairs (α, a) with α in F, a in $\mathfrak{A}$; addition and multiplication by scalars are defined componentwise; multiplication is defined by

$$(\alpha, a)(\beta, b) = (\alpha\beta, \beta a + \alpha b + ab), \qquad \alpha, \beta \text{ in } F, \quad a, b \text{ in } \mathfrak{A}.$$

Then $\mathfrak{A}_1$ is an algebra over F with identity element $1 = (1, 0)$. The set $\mathfrak{A}'$ of all pairs $(0, a)$ in $\mathfrak{A}_1$ with a in $\mathfrak{A}$ is an ideal of $\mathfrak{A}_1$ which is isomorphic to $\mathfrak{A}$. As a vector space $\mathfrak{A}_1$ is the direct sum of $\mathfrak{A}'$ and the 1-dimensional space $F1 = \{\alpha 1 \mid \alpha \text{ in } F\}$. Identifying $\mathfrak{A}'$ with its isomorphic image $\mathfrak{A}$, we can write every element of $\mathfrak{A}_1$ uniquely in the form $\alpha 1 + a$ with α in F, a in $\mathfrak{A}$, in which case the multiplication becomes

$$(\alpha 1 + a)(\beta 1 + b) = (\alpha\beta)1 + (\beta a + \alpha b + ab).$$

We say that we have *adjoined an identity element* to $\mathfrak{A}$ to obtain $\mathfrak{A}_1$. (If $\mathfrak{A}$ is associative, this familiar construction yields an associative algebra $\mathfrak{A}_1$ with 1. A similar statement is readily verifiable for (commutative) Jordan algebras and for alternative algebras. It is of course not true for Lie algebras, since $1^2 = 1 \neq 0$.)

Let $\mathfrak{B}$ and $\mathfrak{A}$ be algebras over a field F. The *Kronecker product* $\mathfrak{B} \otimes_F \mathfrak{A}$ (written $\mathfrak{B} \otimes \mathfrak{A}$ if there is no ambiguity) is the tensor product $\mathfrak{B} \otimes_F \mathfrak{A}$ of the vector spaces $\mathfrak{B}$, $\mathfrak{A}$ (so that all elements are sums $\Sigma b \otimes a$, b in $\mathfrak{B}$, a in $\mathfrak{A}$), multiplication being defined by distributivity and

$$(b_1 \otimes a_1)(b_2 \otimes a_2) = (b_1 b_2) \otimes (a_1 a_2), \qquad b_i \text{ in } \mathfrak{B}, \quad a_i \text{ in } \mathfrak{A}.$$

If $\mathfrak{B}$ contains 1, then the set of all $1 \otimes a$ in $\mathfrak{B} \otimes \mathfrak{A}$ is a subalgebra of $\mathfrak{B} \otimes \mathfrak{A}$ which is isomorphic to $\mathfrak{A}$, and which we can identify with $\mathfrak{A}$ (similarly, if $\mathfrak{A}$ contains 1, then $\mathfrak{B} \otimes \mathfrak{A}$ contains $\mathfrak{B}$ as a subalgebra). If $\mathfrak{B}$ and $\mathfrak{A}$ are finite-dimensional over F, then $\dim(\mathfrak{B} \otimes \mathfrak{A}) = (\dim \mathfrak{B})(\dim \mathfrak{A})$.

We shall on numerous occasions be concerned with the case where $\mathfrak{B}$ is taken to be a field (an arbitrary extension K of F). Then K does contain 1, so $\mathfrak{A}_K = K \otimes_F \mathfrak{A}$ contains $\mathfrak{A}$ (in the sense of isomorphism) as a subalgebra over F. Moreover, $\mathfrak{A}_K$ is readily seen to be an algebra over K, which is called the *scalar extension* of $\mathfrak{A}$ to an algebra over K. The properties of a tensor product ensure that any basis for $\mathfrak{A}$ over F is a basis for $\mathfrak{A}_K$ over K. In case $\mathfrak{A}$ is finite-dimensional over F, this gives an easy representation for the elements of $\mathfrak{A}_K$. Let $u_1, \ldots, u_n$ be any basis for $\mathfrak{A}$ over F. Then the elements of $\mathfrak{A}_K$ are the linear combinations

$$(2.2) \qquad \sum \alpha_i u_i \quad \left(= \sum \alpha_i \otimes u_i \right), \qquad \alpha_i \text{ in } K,$$

where the coefficients α_i in (2.2) are uniquely determined. Addition and multiplication by scalars are performed componentwise. For multiplication in $\mathfrak{A}_K$ we use bilinearity and the multiplication table

$$(2.3) \qquad u_i u_j = \sum \gamma_{ijk} u_k, \qquad \gamma_{ijk} \text{ in } F.$$

The elements of $\mathfrak{A}$ are obtained by restricting the α_i in (2.2) to elements of F.

For finite-dimensional $\mathfrak{A}$, the scalar extension $\mathfrak{A}_K$ (K an arbitrary extension of F) may be defined in a noninvariant way (without recourse to tensor products) by use of a basis as above. Let $u_1, \ldots, u_n$ be any basis for $\mathfrak{A}$ over F; multiplication in $\mathfrak{A}$ is given by the multiplication table

(2.3). Let $\mathfrak{A}_K$ be an n-dimensional algebra over K with the same multiplication table (this is valid since the γ_{ijk}, being in F, are in K). What remains to be verified is that a different choice of basis for $\mathfrak{A}$ over F would yield an algebra isomorphic (over K) to this one. (A noninvariant definition of the Kronecker product of two finite-dimensional algebras $\mathfrak{A}$, $\mathfrak{B}$ may be similarly given.)

For the classes of algebras mentioned in the Introduction (Jordan algebras of characteristic $\neq 2$, and Lie and alternative algebras of arbitrary characteristic), one may verify that algebras remain in the same class under scalar extension—a property which is not shared by classes of algebras defined by more general identities (as, for example, in Chapter V).

Just as the *commutator* $[x, y] = xy - yx$ measures commutativity (and lack of it) in an algebra $\mathfrak{A}$, the *associator*

$$(x, y, z) = (xy)z - x(yz)$$

of any three elements may be introduced as a measure of associativity (and lack of it) in $\mathfrak{A}$. Thus the definitions of alternative and Jordan algebras may be written as

$$(x, x, y) = (y, x, x) = 0 \qquad \text{for all} \quad x, y \text{ in } \mathfrak{A}$$

and

$$[x, y] = (x, y, x^2) = 0 \qquad \text{for all} \quad x, y \text{ in } \mathfrak{A}.$$

Note that the associator (x, y, z) is linear in each argument. One identity which is sometimes useful and which holds in any algebra $\mathfrak{A}$ is

$$(2.4) \qquad a(x, y, z) + (a, x, y)z = (ax, y, z) - (a, xy, z) + (a, x, yz)$$

for all a, x, y, z in $\mathfrak{A}$.

The *nucleus* $\mathfrak{G}$ of an algebra $\mathfrak{A}$ is the set of elements g in $\mathfrak{A}$ which associate with every pair of elements x, y in $\mathfrak{A}$ in the sense that

$$(g, x, y) = (x, g, y) = (x, y, g) = 0 \qquad \text{for all} \quad x, y \text{ in } \mathfrak{A}.$$

It is easy to verify that $\mathfrak{G}$ is an associative subalgebra of $\mathfrak{A}$. $\mathfrak{G}$ is a subspace by the linearity of the associator in each argument, and

$$(g_1 g_2, x, y) = g_1(g_2, x, y) + (g_1, g_2, x)y$$
$$+ (g_1, g_2 x, y) - (g_1, g_2, xy) = 0,$$

etc., by (2.4).

The *center* $\mathfrak{C}$ of $\mathfrak{A}$ is the set of all c in $\mathfrak{A}$ which commute and associate with all elements; that is, the set of all c in the nucleus $\mathfrak{G}$ with the additional property that

$$xc = cx \qquad \text{for all} \quad x \text{ in } \mathfrak{A}.$$

This clearly generalizes the familiar notion of the center of an associative algebra. Note that $\mathfrak{C}$ is a commutative associative subalgebra of $\mathfrak{A}$.

2.   THE ASSOCIATIVE MULTIPLICATION ALGEBRA $\mathfrak{M}(\mathfrak{A})$

Let a be any element of an algebra $\mathfrak{A}$ over F. The *right multiplication* R_a of $\mathfrak{A}$ which is determined by a is defined by

$$R_a : x \to xa \qquad \text{for all} \quad x \text{ in } \mathfrak{A}.$$

Clearly R_a is a linear operator on $\mathfrak{A}$. Also the set $R(\mathfrak{A})$ of all right multiplications of $\mathfrak{A}$ is a subspace of the associative algebra $\mathfrak{E}$ of all linear operators on $\mathfrak{A}$, since $a \to R_a$ is a linear mapping of $\mathfrak{A}$ into $\mathfrak{E}$. (In the familiar case of an associative algebra, $R(\mathfrak{A})$ is a subalgebra of $\mathfrak{E}$, but this is not true in general.) Similarly, the *left multiplication* L_a defined by

$$L_a : x \to ax \qquad \text{for all} \quad x \text{ in } \mathfrak{A}$$

is a linear operator on $\mathfrak{A}$, the mapping $a \to L_a$ is linear, and the set $L(\mathfrak{A})$ of all left multiplications of $\mathfrak{A}$ is a subspace of $\mathfrak{E}$.

We denote by $\mathfrak{M}(\mathfrak{A})$, or simply $\mathfrak{M}$, the enveloping algebra of $R(\mathfrak{A}) \cup L(\mathfrak{A})$; that is, the (associative) subalgebra of $\mathfrak{E}$ generated by right and left multiplications of $\mathfrak{A}$. $\mathfrak{M}(\mathfrak{A})$ is the intersection of all subalgebras of $\mathfrak{E}$ which contain both $R(\mathfrak{A})$ and $L(\mathfrak{A})$. The elements of $\mathfrak{M}(\mathfrak{A})$ are of the form $\sum S_1 \cdots S_h$ where S_i is either a right or a left multiplication of $\mathfrak{A}$. We call the associative algebra $\mathfrak{M} = \mathfrak{M}(\mathfrak{A})$ the *multiplication algebra* of $\mathfrak{A}$.

It is sometimes useful to have a notation for the enveloping algebra of the right and left multiplications (of $\mathfrak{A}$) which correspond to the elements of any subset $\mathfrak{B}$ of $\mathfrak{A}$; we shall write $\mathfrak{B}^*$ for this subalgebra of $\mathfrak{M}(\mathfrak{A})$. That is, $\mathfrak{B}^*$ is the set of all $\sum S_1 \cdots S_h$ where S_i is either R_{b_i}, the right multiplication of $\mathfrak{A}$ determined by b_i in $\mathfrak{B}$, or L_{b_i}. Clearly $\mathfrak{A}^* = \mathfrak{M}(\mathfrak{A})$, but note the difference between $\mathfrak{B}^*$ and $\mathfrak{M}(\mathfrak{B})$ in case $\mathfrak{B}$ is a proper subalgebra of $\mathfrak{A}$—they are associative algebras of operators on different spaces ($\mathfrak{A}$ and $\mathfrak{B}$, respectively).

An algebra $\mathfrak{A}$ over F is called *simple* in case 0 and $\mathfrak{A}$ itself are the only ideals of $\mathfrak{A}$, and $\mathfrak{A}$ is not a zero algebra (equivalently, in the presence of the first assumption, $\mathfrak{A}$ is not the zero algebra of dimension 1). Since an ideal of $\mathfrak{A}$ is an invariant subspace under $\mathfrak{M} = \mathfrak{M}(\mathfrak{A})$, and conversely, it follows that $\mathfrak{A}$ is simple if and only if $\mathfrak{M} \neq 0$ is an irreducible set of linear operators on $\mathfrak{A}$. Since $\mathfrak{A}^2$ ($= \mathfrak{A}\mathfrak{A}$) is an ideal of $\mathfrak{A}$, we have $\mathfrak{A}^2 = \mathfrak{A}$ in case $\mathfrak{A}$ is simple.

An algebra $\mathfrak{A}$ over F is a *division algebra* in case $\mathfrak{A} \neq 0$ and the equations

$$(2.5) \qquad ax = b, \qquad ya = b \qquad (a \neq 0, \quad b \text{ in } \mathfrak{A})$$

have unique solutions x, y in $\mathfrak{A}$; this is equivalent to saying that, for any $a \neq 0$ in $\mathfrak{A}$, L_a and R_a have inverses L_a^{-1} and R_a^{-1}. Any division algebra is simple. For, if $\mathfrak{B} \neq 0$ is merely a left ideal of $\mathfrak{A}$, there is an element $a \neq 0$ in $\mathfrak{B}$ and $\mathfrak{A} \subseteq \mathfrak{A}a \subseteq \mathfrak{B}$ by (2.5), or $\mathfrak{B} = \mathfrak{A}$; also clearly $\mathfrak{A}^2 \neq 0$. [Any associative division algebra $\mathfrak{A}$ has an identity 1, since (2.5) implies that the nonzero elements form a multiplicative group. In general, a division algebra need not contain an identity 1.] If $\mathfrak{A}$ has finite dimension $n \geq 1$ over F, then $\mathfrak{A}$ is a division algebra if and only if $\mathfrak{A}$ is *without zero divisors* ($x \neq 0$ and $y \neq 0$ in $\mathfrak{A}$ imply $xy \neq 0$), inasmuch as the finite-dimensionality ensures that L_a (and similarly R_a), being (1-1) for $a \neq 0$, has an inverse.

In order to make the observation that any simple ring is actually an algebra, so the study of simple rings reduces to that of (possibly infinite-dimensional) simple algebras, we take for granted that the appropriate definitions for rings are apparent, and we digress to consider any simple ring $\mathfrak{R}$. The (associative) multiplication ring $\mathfrak{M} = \mathfrak{M}(\mathfrak{R}) \neq 0$ is irreducible as a ring of endomorphisms of $\mathfrak{R}$. Thus by Schur's lemma the centralizer $\mathfrak{C}'$ of $\mathfrak{M}$ in the ring $\mathfrak{E}$ of all endomorphisms of $\mathfrak{R}$ is an associative division ring. Since $\mathfrak{M}$ is generated by left and right multiplications of $\mathfrak{R}$, $\mathfrak{C}'$ consists of those endomorphisms T in $\mathfrak{E}$ satisfying $R_y T = T R_y$, $L_x T = T L_x$, or

$$(2.6) \qquad (xy)T = (xT)y = x(yT) \qquad \text{for all} \quad x, y \text{ in } \mathfrak{R}.$$

Hence S, T in $\mathfrak{C}'$ imply

$$(xy)ST = ((xS)y)T = (xS)(yT) = (x(yT))S = (xy)TS,$$

so that $zST = zTS$ for all z in $\Re^2 = \Re$. That is, $ST = TS$ for all S, T in $\mathfrak{C}'$; $\mathfrak{C}'$ is a field which we call the *multiplication centralizer* of $\Re$. Now the simple ring $\Re$ may be regarded in a natural way as an algebra over the field $\mathfrak{C}'$. Denote T in $\mathfrak{C}'$ by α, and write $\alpha x = xT$ for any x in $\Re$. Then $\Re$ is a (left) vector space over $\mathfrak{C}'$. Also (2.6) gives the defining relations $\alpha(xy) = (\alpha x)y = x(\alpha y)$ for an algebra over $\mathfrak{C}'$. As an algebra over $\mathfrak{C}'$ (or any subfield F of $\mathfrak{C}'$), $\Re$ is simple since any ideal of $\Re$ as an algebra is *a fortiori* an ideal of $\Re$ as a ring.

Returning now to any simple algebra $\mathfrak{A}$ over F, we recall that the multiplication algebra $\mathfrak{M}(\mathfrak{A})$ is irreducible as a set of linear operators on the vector space $\mathfrak{A}$ over F. But (Jacobson [24], p. 274) this means that $\mathfrak{M}(\mathfrak{A})$ is irreducible as a set of endomorphisms of the additive group of $\mathfrak{A}$, so that $\mathfrak{A}$ is a simple ring. That is, the notions of simple algebra and simple ring coincide.

An algebra $\mathfrak{A}$ over F is called *central simple* in case $\mathfrak{A}_K$ is simple for every extension K of F. Every central simple algebra is simple (take $K = F$). It is proved in Jacobson [25], pp. 290–293, that any simple algebra $\mathfrak{A}$ (of arbitrary dimension), regarded as an algebra over its multiplication centralizer $\mathfrak{C}'$ (so that $\mathfrak{C}' = F$), is central simple. It is also proved there that, if the simple algebra $\mathfrak{A}$ is finite-dimensional over $\mathfrak{C}'$, then $\mathfrak{M}$ is the algebra of all linear operators on $\mathfrak{A}$ over $\mathfrak{C}'$; hence $\mathfrak{C}'$ is the center of $\mathfrak{M}$ in this case.

Theorem 2.1 (Jacobson). *The center $\mathfrak{C}$ of any simple algebra $\mathfrak{A}$ over F is either 0 or a field. In the latter case $\mathfrak{A}$ contains 1, the multiplication centralizer $\mathfrak{C}' = \mathfrak{C}^* = \{R_c \mid c \in \mathfrak{C}\}$, and $\mathfrak{A}$ is a central simple algebra over $\mathfrak{C}$.*

Proof. Note that c is in the center of any algebra $\mathfrak{A}$ if and only if

$$R_c = L_c \qquad \text{and} \qquad [L_c, R_y] = R_c R_y - R_{cy} = R_y R_c - R_{yc} = 0$$

for all y in $\mathfrak{A}$ or, more compactly,

(2.7) $R_c = L_c, \qquad R_c R_y = R_y R_c = R_{cy} \qquad$ for all y in $\mathfrak{A}$.

Hence (2.6) implies that

$$cT \quad \text{is in} \quad \mathfrak{C} \qquad \text{for all} \quad c \text{ in } \mathfrak{C}, \quad T \text{ in } \mathfrak{C}'.$$

For (2.6) may be written as

$$R_y T = TR_y = R_{yT} \qquad \text{for all} \quad y \text{ in } \mathfrak{A},$$

or, equivalently, as

$$L_x T = L_{xT} = TL_x \qquad \text{for all} \quad x \text{ in } \mathfrak{A}.$$

These imply

$$R_{cT} = TR_c = TL_c = L_{cT},$$

together with

$$R_{cT} R_y = R_c TR_y = R_c R_{yT} = R_{c(yT)} = R_{(cT)y}$$

and

$$R_y R_{cT} = R_y R_c T = R_c R_y T = R_c TR_y = R_{cT} R_y.$$

That is, cT is in $\mathfrak{C}$. Note also that (2.7) implies

(2.8) $L_x R_c = R_c L_x \qquad \text{for all} \quad c \text{ in } \mathfrak{C}, \qquad x \text{ in } \mathfrak{A}.$

Since $R_{c_1} R_{c_2} = R_{c_1 c_2}$ (c_i in $\mathfrak{C}$) by (2.7), the subalgebra $\mathfrak{C}^*$ of $\mathfrak{M}(\mathfrak{A})$ is just $\mathfrak{C}^* = \{R_c \mid c \in \mathfrak{C}\}$, and the mapping $c \to R_c$ is a homomorphism of $\mathfrak{C}$ onto $\mathfrak{C}^*$. Also (2.7) and (2.8) imply that R_c commutes with every element of $\mathfrak{M}$ so that $\mathfrak{C}^* \subseteq \mathfrak{C}'$. Moreover, $\mathfrak{C}^*$ is an ideal of the (commutative) field $\mathfrak{C}'$ since $TR_c = R_{cT}$ is in $\mathfrak{C}^*$ for all T in $\mathfrak{C}'$, c in $\mathfrak{C}$. Hence either $\mathfrak{C}^* = 0$ or $\mathfrak{C}^* = \mathfrak{C}'$.

Now $\mathfrak{C}^* = 0$ implies $R_c = 0$ for all c in $\mathfrak{C}$; hence $\mathfrak{C} = 0$. For, if there is $c \neq 0$ in $\mathfrak{C}$, then $\mathfrak{B} = Fc \neq 0$ is an ideal of $\mathfrak{A}$ since $\mathfrak{B}\mathfrak{A} = \mathfrak{A}\mathfrak{B} = 0$. Then $\mathfrak{B} = \mathfrak{A}$, $\mathfrak{A}^2 = 0$, a contradiction.

In the remaining case $\mathfrak{C}^* = \mathfrak{C}'$, the identity operator $1_{\mathfrak{A}}$ on $\mathfrak{A}$ is in $\mathfrak{C}' = \mathfrak{C}^*$. Hence there is an element e in $\mathfrak{C}$ such that $R_e = L_e = 1_{\mathfrak{A}}$, or $ae = ea = a$ for all a in $\mathfrak{A}$; $\mathfrak{A}$ has a unity element $1 = e$. Then $c \to R_c$ is an isomorphism between $\mathfrak{C}$ and the field $\mathfrak{C}'$. $\mathfrak{A}$ is an algebra over the field $\mathfrak{C}$, and as such is central simple.

For any algebra $\mathfrak{A}$ over F, one obtains a *derived series* of subalgebras $\mathfrak{A}^{(1)} \supseteq \mathfrak{A}^{(2)} \supseteq \mathfrak{A}^{(3)} \supseteq \cdots$ by defining $\mathfrak{A}^{(1)} = \mathfrak{A}$, $\mathfrak{A}^{(i+1)} = (\mathfrak{A}^{(i)})^2$. $\mathfrak{A}$ is called *solvable* in case $\mathfrak{A}^{(r)} = 0$ for some integer r.

Proposition 2.2. *If an algebra* $\mathfrak{A}$ *contains a solvable ideal* $\mathfrak{B}$, *and if* $\bar{\mathfrak{A}} = \mathfrak{A}/\mathfrak{B}$ *is solvable, then* $\mathfrak{A}$ *is solvable.*

Proof. Since the natural mapping $a \to \bar{a} = a + \mathfrak{B}$ (a in $\mathfrak{A}$, $\bar{a}$ in $\mathfrak{A}/\mathfrak{B}$) is a homomorphism, it follows that $\overline{\mathfrak{A}^2} = \bar{\mathfrak{A}}^2$ and that $\overline{\mathfrak{A}^{(i)}} = \bar{\mathfrak{A}}^{(i)}$. Then $\bar{\mathfrak{A}}^{(r)} = 0$ implies $\overline{\mathfrak{A}^{(r)}} = 0$, or $\mathfrak{A}^{(r)} \subseteq \mathfrak{B}$. But $\mathfrak{B}^{(s)} = 0$ for some s, so $\mathfrak{A}^{(r+s)} = (\mathfrak{A}^{(r)})^{(s)} \subseteq \mathfrak{B}^{(s)} = 0$. Hence $\mathfrak{A}$ is solvable.

Proposition 2.3. *If* $\mathfrak{B}$ *and* $\mathfrak{C}$ *are solvable ideals of an algebra* $\mathfrak{A}$, *then* $\mathfrak{B} + \mathfrak{C}$ *is a solvable ideal of* $\mathfrak{A}$. *Hence, if* $\mathfrak{A}$ *is finite-dimensional,* $\mathfrak{A}$ *has a unique maximal solvable ideal* $\mathfrak{N}$. *Moreover, the only solvable ideal of* $\mathfrak{A}/\mathfrak{N}$ *is* 0.

Proof. $\mathfrak{B} + \mathfrak{C}$ is an ideal because $\mathfrak{B}$ and $\mathfrak{C}$ are ideals. By the second isomorphism theorem $(\mathfrak{B} + \mathfrak{C})/\mathfrak{C} \cong \mathfrak{B}/(\mathfrak{B} \cap \mathfrak{C})$. But $\mathfrak{B}/(\mathfrak{B} \cap \mathfrak{C})$ is a homomorphic image of the solvable algebra $\mathfrak{B}$, and is therefore clearly solvable. Then $\mathfrak{B} + \mathfrak{C}$ is solvable by Proposition 2.2. It follows that, if $\mathfrak{A}$ is finite-dimensional, the solvable ideal of maximum dimension is unique (and contains every solvable ideal of $\mathfrak{A}$). Let $\mathfrak{N}$ be this maximal solvable ideal, and $\bar{\mathfrak{G}}$ be any solvable ideal of $\bar{\mathfrak{A}} = \mathfrak{A}/\mathfrak{N}$. The complete inverse image $\mathfrak{G}$ of $\bar{\mathfrak{G}}$ under the natural homomorphism of $\mathfrak{A}$ onto $\bar{\mathfrak{A}}$ is an ideal of $\mathfrak{A}$ such that $\mathfrak{G}/\mathfrak{N} = \bar{\mathfrak{G}}$. Then $\mathfrak{G}$ is solvable by Proposition 2.2, so $\mathfrak{G} \subseteq \mathfrak{N}$. Hence $\mathfrak{G}/\mathfrak{N} = \bar{\mathfrak{G}} = 0$.

An algebra $\mathfrak{A}$ is called *nilpotent* in case there exists an integer t such that any product $z_1 z_2 \cdots z_t$ of t elements in $\mathfrak{A}$, no matter how associated, is 0. This clearly generalizes the concept of nilpotence as defined for associative algebras. Also any nilpotent algebra is solvable.

Theorem 2.4. *An ideal* $\mathfrak{B}$ *of an algebra* $\mathfrak{A}$ *is nilpotent if and only if the (associative) subalgebra* $\mathfrak{B}^*$ *of* $\mathfrak{M}(\mathfrak{A})$ *is nilpotent.*

Proof. Suppose that every product of t elements of $\mathfrak{B}$, no matter how associated, is 0. Then the same is true for any product of more than t elements of $\mathfrak{B}$. Let $T = T_1 \cdots T_t$ be any product of t elements of $\mathfrak{B}^*$. Then T is a sum of terms each of which is a product of at least t linear operators S_i, each S_i being either L_{b_i} or R_{b_i} (b_i in $\mathfrak{B}$). Since $\mathfrak{B}$ is an ideal of $\mathfrak{A}$, $x S_1$ is in $\mathfrak{B}$ for every x in $\mathfrak{A}$. Hence xT is a sum of terms, each of which is a product of at least t elements in $\mathfrak{B}$. Hence $xT = 0$ for all x in $\mathfrak{A}$, or $T = 0$, $\mathfrak{B}^*$ is nilpotent. For the converse we need only that $\mathfrak{B}$ is a subalgebra of $\mathfrak{A}$. We show by induction on n that any product of at least 2^n elements in $\mathfrak{B}$, no matter how associated, is of the form $bS_1 \cdots S_n$ with b in $\mathfrak{B}$, S_i in $\mathfrak{B}^*$. For $n = 1$, we take any product of at least 2 elements in $\mathfrak{B}$. There is a final multiplication which is performed. Since $\mathfrak{B}$ is a subalgebra, each of the two factors is in $\mathfrak{B}$: $bb_1 = bR_{b_1} = bS_1$. Similarly in any product of at least 2^{n+1} elements of $\mathfrak{B}$, no matter how associated, there is a final multiplication which is performed. At least one of the two factors is a product of at least 2^n elements of $\mathfrak{B}$, while the other factor b' is in $\mathfrak{B}$. Hence by the assumption of the induction we have either

$$b'(bS_1 \cdots S_n) = bS_1 \cdots S_n L_{b'} = bS_1 \cdots S_{n+1}$$

or

$$(bS_1 \cdots S_n)b' = bS_1 \cdots S_n R_{b'} = bS_1 \cdots S_{n+1},$$

as desired. Hence, if any product $S_1 \cdots S_t$ of t elements in $\mathfrak{B}^*$ is 0, any product of 2^t elements of $\mathfrak{B}$, no matter how associated, is 0. That is, $\mathfrak{B}$ is nilpotent.

3. THE LIE MULTIPLICATION ALGEBRA $\mathfrak{L}(\mathfrak{A})$

We have seen how the (associative) multiplication algebra $\mathfrak{M}(\mathfrak{A})$ may be employed in studying an arbitrary algebra $\mathfrak{A}$. The known theory of Lie algebras may similarly be used in investigating nonassociative algebras, not only through the derivation algebras mentioned in the Introduction, but also through the Lie analogue of $\mathfrak{M}(\mathfrak{A})$.

Let $\mathfrak{H}$ be a subspace of $\mathfrak{E} = \mathfrak{E}(V)$, the associative algebra of all linear operators on some vector space V over F. By the Lie enveloping algebra of $\mathfrak{H}$ we mean the (Lie) subalgebra of $\mathfrak{E}^-$ generated by $\mathfrak{H}$; that is, the intersection of all subalgebras of $\mathfrak{E}^-$ containing $\mathfrak{H}$. Defining $\mathfrak{H}_i$ inductively by $\mathfrak{H}_1 = \mathfrak{H}$, $\mathfrak{H}_{i+1} = [\mathfrak{H}_1, \mathfrak{H}_i]$, we see that the Lie enveloping algebra of $\mathfrak{H}$ is

$$\sum_{i=1}^{\infty} \mathfrak{H}_i = \mathfrak{H}_1 + \mathfrak{H}_2 + \mathfrak{H}_3 + \cdots.$$

For

(2.9) $[\mathfrak{H}_i, \mathfrak{H}_j] \subseteq \mathfrak{H}_{i+j}$ $(i, j = 1, 2, 3, \ldots),$

as may be seen from anticommutativity and the Jacobi identity. The definition $\mathfrak{H}_{j+1} = [\mathfrak{H}_1, \mathfrak{H}_j]$ gives the case $i = 1$. The Jacobi identity implies that

$$(2.10) \qquad [[\mathfrak{A}, \mathfrak{B}], \mathfrak{C}] \subseteq [[\mathfrak{B}, \mathfrak{C}], \mathfrak{A}] + [[\mathfrak{C}, \mathfrak{A}], \mathfrak{B}]$$

for subspaces $\mathfrak{A}, \mathfrak{B}, \mathfrak{C}$, of $\mathfrak{C}^-$. By induction on i, we assume (2.9) for all j and have

$$[\mathfrak{H}_{i+1}, \mathfrak{H}_j] = [[\mathfrak{H}_1, \mathfrak{H}_i], \mathfrak{H}_j] \subseteq [[\mathfrak{H}_i, \mathfrak{H}_j], \mathfrak{H}_1] + [[\mathfrak{H}_j, \mathfrak{H}_1], \mathfrak{H}_i]$$

$$\subseteq [\mathfrak{H}_{i+j}, \mathfrak{H}_1] + [\mathfrak{H}_{j+1}, \mathfrak{H}_i]$$

$$= [\mathfrak{H}_1, \mathfrak{H}_{i+j}] + [\mathfrak{H}_i, \mathfrak{H}_{j+1}] \subseteq \mathfrak{H}_{i+j+1}.$$

It follows from (2.9) that $\sum_{i=1}^{\infty} \mathfrak{H}_i$ is a Lie subalgebra of $\mathfrak{C}^-$. Since the Lie enveloping algebra of $\mathfrak{H}$ contains $\sum_{i=1}^{\infty} \mathfrak{H}_i$, we have the desired equality.

Let $\mathfrak{A}$ be an algebra over F. We define the *Lie multiplication algebra* $\mathfrak{L} = \mathfrak{L}(\mathfrak{A})$ of $\mathfrak{A}$ to be the Lie enveloping algebra of $\mathfrak{H} = R(\mathfrak{A}) + L(\mathfrak{A})$; that is, $\mathfrak{L}$ is the Lie algebra (subalgebra of $\mathfrak{C}^-$) generated by the right and left multiplications of $\mathfrak{A}$. Clearly $\mathfrak{L} \subseteq \mathfrak{M}(\mathfrak{A})$. Also the center $\mathfrak{Z}$ of $\mathfrak{L}$ is contained in the center of $\mathfrak{M}(\mathfrak{A})$, for every element of $\mathfrak{Z}$ commutes with every element of $\mathfrak{H}$ (which generates $\mathfrak{M}(\mathfrak{A})$).

We recall that a derivation D of $\mathfrak{A}$ is a linear operator on $\mathfrak{A}$ satisfying

$$(xy)D = (xD)y + x(yD) \qquad \text{for all} \quad x, y \text{ in } \mathfrak{A}.$$

This may be expressed in terms of right or left multiplications of $\mathfrak{A}$: D in $\mathfrak{C}$ is a derivation of $\mathfrak{A}$ if and only if

$$[R_y, D] = R_{yD} \qquad \text{for all} \quad y \text{ in } \mathfrak{A};$$

equivalently, D in $\mathfrak{C}$ is a derivation if and only if

$$[L_x, D] = L_{xD} \qquad \text{for all} \quad x \text{ in } \mathfrak{A}.$$

If $\mathfrak{A}$ contains 1, then

$$1D = 0 \qquad \text{for all} \quad D \text{ in } \mathfrak{D}(\mathfrak{A}),$$

since $1D = (11)D = (1D)1 + 1(1D) = 2(1D)$.

A subalgebra $\mathfrak{S}$ of $\mathfrak{A}$ is called *characteristic* in case $\mathfrak{S}\mathfrak{D}(\mathfrak{A}) \subseteq \mathfrak{S}$. The center $\mathfrak{C}$ of $\mathfrak{A}$ is characteristic, for (2.7) implies that

$$R_{cD} = [R_c, D] = [L_c, D] = L_{cD},$$

$$R_{cD}R_y = [R_c, D]R_y = [R_c R_y, D] - R_c[R_y, D]$$

$$= [R_y R_c, D] - [R_y, D]R_c = R_y[R_c, D]$$

$$= R_y R_{cD}$$

and also

$$R_{cD}R_y = [R_y R_c, D] - [R_y, D]R_c = [R_{cy}, D] - R_c R_{yD}$$

$$= R_{(cy)D - c(yD)} = R_{(cD)y}.$$

We shall call a derivation D of $\mathfrak{A}$ *inner* in case D is in $\mathfrak{L} = \mathfrak{L}(\mathfrak{A})$, the Lie multiplication algebra of $\mathfrak{A}$. The set $\mathfrak{J} = \mathfrak{L} \cap \mathfrak{D}(\mathfrak{A})$ of all inner derivations of $\mathfrak{A}$ is an ideal of $\mathfrak{D}(\mathfrak{A})$, since $[\mathfrak{H}, \mathfrak{D}(\mathfrak{A})] \subseteq \mathfrak{H}$ where $\mathfrak{H} = R(\mathfrak{A}) + L(\mathfrak{A})$. This is the case $i = 1$ of $[\mathfrak{H}_i, \mathfrak{D}(\mathfrak{A})] \subseteq \mathfrak{H}_i$ which follows from (2.10) by induction. Hence $\mathfrak{L} = \sum \mathfrak{H}_i$ implies $[\mathfrak{L}, \mathfrak{D}(\mathfrak{A})] \subseteq \mathfrak{L}$, whence $[\mathfrak{J}, \mathfrak{D}(\mathfrak{A})] \subseteq \mathfrak{L} \cap \mathfrak{D}(\mathfrak{A}) = \mathfrak{J}$. Similarly, we have $[\mathfrak{M}(\mathfrak{A}), \mathfrak{D}(\mathfrak{A})] \subseteq \mathfrak{M}(\mathfrak{A})$.

Our definition of inner derivation for a nonassociative algebra agrees with the usual definition for Lie algebras, and with that for associative algebras having a left (or right) identity; for example, for finite-dimensional semisimple associative algebras.

For, if $\mathfrak{A}$ is a Lie algebra, then $L_x = - R_x$ by anticommutativity, while the Jacobi identity implies that $[R_x, R_y] = R_{xy}$. Hence $\mathfrak{H} = R(\mathfrak{A})$ and $[R(\mathfrak{A}), R(\mathfrak{A})] \subseteq R(\mathfrak{A})$ so that, if $\mathfrak{A}$ is a Lie algebra, we have $\mathfrak{L} = R(\mathfrak{A})$. Then D is in $\mathfrak{L}$ (and is an inner derivation of $\mathfrak{A}$) if and only if $D = R_x$ ($= \operatorname{ad} x$) for some x in $\mathfrak{A}$. On the other hand, if $\mathfrak{A}$ is associative, then

$$R_x R_y = R_{xy}, \qquad R_x L_y = L_y R_x, \qquad \text{and} \qquad L_x L_y = L_{yx}.$$

These imply that $\mathfrak{L} = R(\mathfrak{A}) + L(\mathfrak{A})$. Hence D is in $\mathfrak{L}$ if and only if $D = R_x + L_y$. If $\mathfrak{A}$ contains a left identity e, then $L_e = 1_{\mathfrak{A}}$ and

$$0 = L_{eD} = L_{ex + ye} = L_x + L_e L_y = L_x + L_y,$$

so that $D = R_x - L_x$ (which is a derivation of $\mathfrak{A}$ for every x in $\mathfrak{A}$).

Guided by the known structure of finite-dimensional associative algebras (of arbitrary characteristic) and Lie algebras of characteristic 0, we study finite-dimensional algebras $\mathfrak{A}$ having the structure

$$\mathfrak{A} = \mathfrak{S}_1 \oplus \cdots \oplus \mathfrak{S}_t$$

for simple ideals $\mathfrak{S}_i$ ($i = 1, \ldots, t$). It is easy to see that the $\mathfrak{S}_i$ are uniquely determined by $\mathfrak{A}$. In an arbitrary algebra $\mathfrak{A}$, $\mathfrak{B}$ is an ideal if and only if $\mathfrak{B}$ is a subspace of $\mathfrak{A}$ which is invariant under the set $\mathfrak{H} = R(\mathfrak{A}) + L(\mathfrak{A})$, and therefore invariant under the associative algebra $\mathfrak{M}(\mathfrak{A})$ and the Lie algebra $\mathfrak{L}(\mathfrak{A})$ generated by $\mathfrak{H}$. Hence, if a finite-dimensional algebra $\mathfrak{A}$ is $\mathfrak{A} = \mathfrak{S}_1 \oplus \cdots \oplus \mathfrak{S}_t$ for simple ideals $\mathfrak{S}_i$, then the set $L(\mathfrak{A}) + R(\mathfrak{A})$ (resp. $\mathfrak{M}(\mathfrak{A})$, $\mathfrak{L}(\mathfrak{A})$) is completely reducible (Jacobson [25], p. 46).

The simple components $\mathfrak{S}_i$ of $\mathfrak{A} = \mathfrak{S}_1 \oplus \cdots \oplus \mathfrak{S}_t$ are characteristic: $\mathfrak{S}_i \mathfrak{D}(\mathfrak{A}) \subseteq \mathfrak{S}_i$ ($i = 1, \ldots, t$). For $\mathfrak{S}_i{}^2 = \mathfrak{S}_i$ implies that x in $\mathfrak{S}_i$ may be written as $x = \sum yz$ (y, z in $\mathfrak{S}_i$). Then $xD = (\sum yz)D = \sum (yD)z + \sum y(zD)$ is in $\mathfrak{S}_i$ for all D in $\mathfrak{D}(\mathfrak{A})$, since $\mathfrak{S}_i$ is an ideal of $\mathfrak{A}$.

We use the known result that every derivation of a finite-dimensional semisimple Lie algebra of characteristic 0 is inner (Jacobson [25], p. 74) to prove a (partial) generalization to arbitrary nonassociative algebras. For the proof of Theorem 2.5 below we require the elementary fact that, if a field K is a finite separable extension of F, then the only derivation of K (over F) is 0 (Zariski and Samuel [1], p. 124).

Theorem 2.5 (Jacobson). *Let $\mathfrak{A}$ be a finite-dimensional algebra which is a direct sum $\mathfrak{A} = \mathfrak{S}_1 \oplus \cdots \oplus \mathfrak{S}_t$ of simple ideals $\mathfrak{S}_i$ over F of characteristic 0, and let $\mathfrak{A}$ contain a left (or right) identity. Then every derivation D of $\mathfrak{A}$ is inner.*

Proof. It is sufficient to prove this for simple algebras $\mathfrak{A}$. For, if $\mathfrak{A} = \mathfrak{S}_1 \oplus \cdots \oplus \mathfrak{S}_t$, we have seen that each $\mathfrak{S}_i$ is characteristic. It follows that $\mathfrak{D}(\mathfrak{A}) = \mathfrak{D}(\mathfrak{S}_1) \oplus \cdots \oplus \mathfrak{D}(\mathfrak{S}_t)$. Similarly, $\mathfrak{L}(\mathfrak{A}) = \mathfrak{L}_1 \oplus \cdots \oplus \mathfrak{L}_t$ where $\mathfrak{L}_i$ is the Lie multiplication algebra of $\mathfrak{S}_i$. It is easy to see from the manner in which these decompositions of $\mathfrak{D}(\mathfrak{A})$ and $\mathfrak{L}(\mathfrak{A})$ are obtained that every derivation of $\mathfrak{A}$ is inner if and only if every derivation of $\mathfrak{S}_i$ is inner ($i = 1, \ldots, t$).

Let D be a derivation of a simple algebra $\mathfrak{A}$. We have remarked that $[\mathfrak{M}(\mathfrak{A}), \mathfrak{D}(\mathfrak{A})] \subseteq \mathfrak{M}(\mathfrak{A})$. Hence the mapping

$$\tilde{D} : T \to [T, D] \qquad \text{for all} \quad T \text{ in } \mathfrak{M}(\mathfrak{A})$$

is a derivation of $\mathfrak{M}(\mathfrak{A})$. The Lie multiplication algebra $\mathfrak{L} = \mathfrak{L}(\mathfrak{A})$ of $\mathfrak{A}$

is completely reducible. Hence $\mathfrak{L} = \mathfrak{L}' \oplus \mathfrak{Z}$ where $\mathfrak{L}' = [\mathfrak{L}, \mathfrak{L}]$ is semisimple and $\mathfrak{Z}$ is the center of $\mathfrak{L}$ (Jacobson [25], p. 47). Since $[\mathfrak{L}, \mathfrak{D}(\mathfrak{A})] \subseteq \mathfrak{L}$, $\tilde{D}$ induces derivations on $\mathfrak{L}$ and $\mathfrak{L}'$. We have remarked before that finite-dimensionality implies that the multiplication centralizer of $\mathfrak{A}$ is the center K of $\mathfrak{M}(\mathfrak{A})$, whence K is a finite (separable) extension of F. We have seen also that $\mathfrak{Z} \subseteq K$. Since $\tilde{D}$ induces a derivation on K, we have $\mathfrak{Z}\tilde{D} \subseteq K\tilde{D} = 0$. Since $\mathfrak{L}'$ is semisimple, the derivation which $\tilde{D}$ induces on $\mathfrak{L}'$ is inner. Since $\mathfrak{Z}\tilde{D} = 0$, this inner derivation of $\mathfrak{L}'$ can be extended to an inner derivation of $\mathfrak{L} = \mathfrak{L}' \oplus \mathfrak{Z}$. Now $R(\mathfrak{A}) + L(\mathfrak{A}) \subseteq \mathfrak{L}$, so there exists $U \in \mathfrak{L}$ satisfying

$$[R_a, D] = R_a \tilde{D} = [R_a, U] \qquad \text{for all} \quad a \text{ in } \mathfrak{A}$$

and

$$[L_a, D] = L_a \tilde{D} = [L_a, U] \qquad \text{for all} \quad a \text{ in } \mathfrak{A}.$$

If there is a left identity element e in $\mathfrak{A}$, then

$$aD = eR_{aD} = e[R_a, D] = e[R_a, U] = aU - (eU)a \qquad \text{for all} \quad a \text{ in } \mathfrak{A},$$

or

$$D = U - L_{eU} \quad \text{in} \quad \mathfrak{L};$$

that is, D is inner. If there is a right identity in $\mathfrak{A}$, it follows similarly that D is inner.

Because of the hypothesis concerning a left (or right) identity, Theorem 2.5 is not a true generalization of the Lie algebra case. Theorem 2.5 does apply, however, to the algebras (alternative, Jordan, etc.) considered in this book. It would be desirable to have either a proof of Theorem 2.5 with the hypothesis concerning a left (or right) identity deleted, or a counterexample.

4. TRACE FORMS; BIMODULES

A very important tool in the study of finite-dimensional Lie algebras $\mathfrak{L}$ of characteristic 0 is the *Killing form*

$$(x, y) = \text{trace } R_x R_y \quad [= \text{trace}(\text{ad } x)(\text{ad } y)] \qquad \text{for all} \quad x, y \text{ in } \mathfrak{L}.$$

The Killing form of $\mathfrak{L}$ is a symmetric bilinear form on $\mathfrak{L}$ which is nondegenerate if and only if $\mathfrak{L}$ is semisimple (that is, the maximal solvable ideal of $\mathfrak{L}$ is 0). The Killing form of $\mathfrak{L}$ is a trace form on $\mathfrak{L}$ according to the following definition.

A symmetric bilinear form (x, y) defined on an arbitrary algebra $\mathfrak{A}$ is called a *trace form* (*associative* or *invariant* symmetric bilinear form) on $\mathfrak{A}$ in case

$$(2.11) \qquad (xy, z) = (x, yz) \qquad \text{for all} \quad x, y, z \text{ in } \mathfrak{A}.$$

If $\mathfrak{B}$ is any ideal of an algebra $\mathfrak{A}$ on which such a bilinear form is defined, then $\mathfrak{B}^{\perp} = \{y \,|\, (x, y) = 0 \text{ for all } x \text{ in } \mathfrak{B}\}$ is also an ideal of $\mathfrak{A}$. For x in $\mathfrak{B}$, y in $\mathfrak{B}^{\perp}$, a in $\mathfrak{A}$ imply that xa and ax are in $\mathfrak{B}$, implying

$$(x, ay) = (xa, y) = 0 \qquad \text{and} \qquad (x, ya) = (ya, x) = (y, ax) = 0.$$

In particular, the radical $\mathfrak{A}^{\perp}$ of the trace form is an ideal of $\mathfrak{A}$.

We also remark that it follows from (2.11) that

$$(xR_y, z) = (x, zL_y) \qquad \text{and} \qquad (xL_y, z) = (z, yx) = (zy, x) = (x, zR_y)$$

so that, for right (or left) multiplications S_i determined by b_i, we have

$$(2.12) \qquad (xS_1 S_2 \cdots S_h, y) = (x, yS_h' \cdots S_2' S_1')$$

where S_i' is the left (or right) multiplication determined by b_i. Hence, if $\mathfrak{B}$ is any subset of $\mathfrak{A}$,

$$(2.13) \qquad (xT, y) = (x, yT') \qquad \text{for all} \quad x, y \text{ in } \mathfrak{A}, \ T \text{ in } \mathfrak{B}^*,$$

where T' may be taken to be in $\mathfrak{B}^*$.

Theorem 2.6 (Dieudonné). *Let $\mathfrak{A}$ be a finite-dimensional algebra over F (of arbitrary characteristic) satisfying*
   (i) *there is a nondegenerate (associative) trace form (x, y) defined on $\mathfrak{A}$, and*
   (ii) $\mathfrak{B}^2 \neq 0$ *for every ideal $\mathfrak{B} \neq 0$ of $\mathfrak{A}$.*
Then $\mathfrak{A}$ is (uniquely) expressible as a direct sum $\mathfrak{A} = \mathfrak{S}_1 \oplus \cdots \oplus \mathfrak{S}_t$ of simple ideals $\mathfrak{S}_i$.

Proof. Let $\mathfrak{S} \, (\neq 0)$ be a minimal ideal of $\mathfrak{A}$. Since (x, y) is a trace form, $\mathfrak{S}^{\perp}$ is an ideal of $\mathfrak{A}$. Hence the intersection $\mathfrak{S} \cap \mathfrak{S}^{\perp}$ is either 0 or

$\mathfrak{S}$, since $\mathfrak{S}$ is minimal. We show that $\mathfrak{S}$ totally isotropic ($\mathfrak{S} \subseteq \mathfrak{S}^{\perp}$) leads to a contradiction.

For, since $\mathfrak{S}^2 \neq 0$ by (ii), we know that the ideal of $\mathfrak{A}$ generated by $\mathfrak{S}^2$ must be the minimal ideal $\mathfrak{S}$. Thus $\mathfrak{S} = \mathfrak{S}^2 + \mathfrak{S}^2\mathfrak{M}$ where $\mathfrak{M}$ is the multiplication algebra of $\mathfrak{A}$. Any element s in $\mathfrak{S}$ may be written in the form $s = \sum (a_i b_i) T_i$ for a_i, b_i in $\mathfrak{S}$, where either $T_i = T_i{}'$ is the identity operator $1_{\mathfrak{A}}$ or T_i is in $\mathfrak{M}$. For every y in $\mathfrak{A}$ we have by (2.13) that

$$(s, y) = \sum((a_i b_i) T_i, y) = \sum (a_i b_i, y T_i{}') = \sum (a_i, b_i(y T_i{}')) = 0$$

since $b_i(y T_i{}')$ is in $\mathfrak{S} \subseteq \mathfrak{S}^{\perp}$. Then $s = 0$ since (x, y) is nondegenerate; $\mathfrak{S} = 0$, a contradiction.

Hence $\mathfrak{S} \cap \mathfrak{S}^{\perp} = 0$; that is, $\mathfrak{S}$ is nonisotropic. Hence $\mathfrak{A} = \mathfrak{S} \perp \mathfrak{S}^{\perp}$ and $\mathfrak{S}^{\perp}$ is nonisotropic (Jacobson [24], p. 151; Artin [2], p. 117). That is, $\mathfrak{A} = \mathfrak{S} \oplus \mathfrak{S}^{\perp}$, the direct sum of ideals $\mathfrak{S}$, $\mathfrak{S}^{\perp}$, and the restriction of (x, y) to $\mathfrak{S}^{\perp}$ is a nondegenerate (associative) trace form defined on $\mathfrak{S}^{\perp}$. That is, (i) holds for $\mathfrak{S}^{\perp}$ as well as $\mathfrak{A}$. Moreover, any ideal of the direct summand $\mathfrak{S}$ or $\mathfrak{S}^{\perp}$ is an ideal of $\mathfrak{A}$; hence $\mathfrak{S}$ is simple and (ii) holds for $\mathfrak{S}^{\perp}$. Induction on the dimension of $\mathfrak{A}$ completes the proof.

We conclude this chapter on arbitrary nonassociative algebras by defining the equivalent notions of bimodule and (bi)representation for any class of nonassociative algebras over F.

Let $\mathscr{C}$ be a class of nonassociative algebras over F. If $\mathfrak{A}$ is in $\mathscr{C}$, and if $\mathfrak{M}$ is a vector space over F, let ma and am in $\mathfrak{M}$ be two bilinear compositions for a in $\mathfrak{A}$, m in $\mathfrak{M}$. Then the direct sum $\mathfrak{A} + \mathfrak{M}$ of the vector spaces $\mathfrak{A}$ and $\mathfrak{M}$ is turned into a nonassociative algebra over F (the *semidirect sum*, or *split null extension*) by defining multiplication in $\mathfrak{A} + \mathfrak{M}$ by

$$(a_1 + m_1)(a_2 + m_2) = a_1 a_2 + (m_1 a_2 + a_1 m_2)$$

for all a_i in $\mathfrak{A}$, m_i in $\mathfrak{M}$. If $\mathfrak{A} + \mathfrak{M}$ is in $\mathscr{C}$, then $\mathfrak{M}$ is called a *bimodule* for $\mathfrak{A}$ in $\mathscr{C}$.

For example, if $\mathscr{C}$ is the class of all nonassociative algebras over F, then no conditions other than the originally assumed bilinearity are imposed on the compositions am and ma. If $\mathscr{C}$ is the class of all associative algebras, then the compositions in $\mathfrak{M}$ must satisfy

$$(ma)b = m(ab), \qquad (am)b = a(mb), \qquad (ab)m = a(bm)$$

for a, b in $\mathfrak{A}$, m in $\mathfrak{M}$; that is, $\mathfrak{M}$ is what is familiarly known as a two-sided $\mathfrak{A}$-module for the associative algebra $\mathfrak{A}$.

If $\mathscr{C}$ is any class of nonassociative algebras defined by multilinear identities $I_i(a_1, a_2, ..., a_{n_i}) = 0$, then $\mathfrak{M}$ is a bimodule for $\mathfrak{A}$ in $\mathscr{C}$ in case all of the identities, obtained by replacing any single a_j in $\mathfrak{A}$ in the identities $I_i(a_1, a_2, ..., a_{n_i}) = 0$ by m in $\mathfrak{M}$, are satisfied (Eilenberg [1]). The defining identities for alternative and Jordan algebras may be linearized (at a slight cost involving the characteristic of F). Bimodules for alternative and Jordan algebras are then easily defined by this procedure.

If $\mathfrak{M}$ is a bimodule for $\mathfrak{A}$, then the mappings $S_a : m \to ma$ and $T_a : m \to am$ are linear operators on $\mathfrak{M}$, and $a \to S_a$, $a \to T_a$ are linear mappings of $\mathfrak{A}$ into the algebra $\mathfrak{E}(\mathfrak{M})$ of all linear operators on $\mathfrak{M}$. A pair (S, T) of linear mappings $a \to S_a$, $a \to T_a$ of $\mathfrak{A}$ into some $\mathfrak{E}(\mathfrak{M})$ is called a *representation* of $\mathfrak{A}$ in $\mathscr{C}$ in case $\mathfrak{M}$, equipped with the compositions $ma = mS_a$, $am = mT_a$, is a bimodule for $\mathfrak{A}$ in $\mathscr{C}$. Clearly the concepts of a representation of $\mathfrak{A}$ and of a bimodule for $\mathfrak{A}$ are equivalent.

Any algebra $\mathfrak{A}$ in a class $\mathscr{C}$ defined by multilinear identities has the regular representation (R, L) where $\mathfrak{M}$ is $\mathfrak{A}$ itself, and R_a and L_a are right and left multiplications in $\mathfrak{A}$. An important simplification is possible if $\mathscr{C}$ contains only commutative (or anticommutative) algebras. For then $ma = am$ (or $ma = -am$) implies $S_a = T_a$ (or $S_a = -T_a$), so that effectively only one mapping $S : a \to S_a$ is involved instead of the pair (S, T). This simplification can be made for Jordan and Lie algebras (for Lie algebras the notion of representation is the usual one).

These ideas have been used, in particular, to generalize the result in Theorem 2.5 for alternative and Jordan algebras. We state these results in Chapters III and IV.

III

ALTERNATIVE ALGEBRAS

1. NILPOTENT ALGEBRAS

As indicated in the Introduction, an *alternative algebra* $\mathfrak{A}$ over F is an algebra in which

$$x^2 y = x(xy), \qquad yx^2 = (yx)x \qquad \text{for all} \quad x, y \text{ in } \mathfrak{A}.$$

The left and right equations are known, respectively, as the *left* and *right alternative laws*. They are equivalent, in terms of associators, to

$$(x, x, y) = (y, x, x) = 0 \qquad \text{for all} \quad x, y \text{ in } \mathfrak{A}$$

or, in terms of left and right multiplications, to

$$(3.1) \qquad L_{x^2} = L_x^2, \qquad R_{x^2} = R_x^2 \qquad \text{for all} \quad x \text{ in } \mathfrak{A}.$$

The associator (x_1, x_2, x_3) "alternates" in the sense that, for any permutation σ of 1, 2, 3 we have $(x_{1\sigma}, x_{2\sigma}, x_{3\sigma}) = (\text{sgn } \sigma)(x_1, x_2, x_3)$. To establish this it is sufficient to prove

$$(x, y, z) = -(y, x, z) = (y, z, x) \qquad \text{for all} \quad x, y, z \text{ in } \mathfrak{A}.$$

Now

$$(x + y, x + y, z) = (x, x, z) + (x, y, z) + (y, x, z) + (y, y, z)$$

$$= (x, y, z) + (y, x, z) = 0,$$

implying $(x, y, z) = -(y, x, z)$. Similarly, $(y, z, x) = -(y, x, z)$.

27

The fact that the associator alternates is equivalent to

$$(3.2) \qquad R_x R_y - R_{xy} = L_{xy} - L_y L_x = L_y R_x - R_x L_y$$
$$= L_x L_y - L_{yx} = R_y L_x - L_x R_y = R_{yx} - R_y R_x$$

for all x, y in $\mathfrak{A}$. It follows from (3.1) and (3.2) that any scalar extension $\mathfrak{A}_K$ of an alternative algebra $\mathfrak{A}$ is alternative.

Our identities on associators imply that in an alternative algebra $\mathfrak{A}$ we have

$$(x, y, x) = 0 \qquad \text{for all} \quad x, y \text{ in } \mathfrak{A};$$

that is,

$$(3.3) \qquad (xy)x = x(yx) \qquad \text{for all} \quad x, y \text{ in } \mathfrak{A},$$

or

$$L_x R_x = R_x L_x \qquad \text{for all} \quad x \text{ in } \mathfrak{A}.$$

Identity (3.3) is called the *flexible* law. All of the algebras mentioned in the Introduction (Lie, Jordan, and alternative) are flexible. The linearized form of the flexible law is

$$(x, y, z) + (z, y, x) = 0 \qquad \text{for all} \quad x, y, z \text{ in } \mathfrak{A}.$$

We shall have occasion to use the Moufang identities

$$(3.4) \qquad\qquad (xax)y = x[a(xy)],$$
$$(3.5) \qquad\qquad y(xax) = [(yx)a]x,$$
$$(3.6) \qquad\qquad (xy)(ax) = x(ya)x$$

for all x, y, a in an alternative algebra $\mathfrak{A}$, where we may write xax unambiguously by (3.3). For

$$
\begin{aligned}
(xax)y - x[a(xy)] &= (xa, x, y) + (x, a, xy) \\
&= -(x, xa, y) - (x, xy, a) \\
&= -[x(xa)]y + x[(xa)y] - [x(xy)]a + x[(xy)a] \\
&= -(x^2 a)y - (x^2 y)a + x[(xa)y + (xy)a] \\
&= -(x^2, a, y) - (x^2, y, a) - x^2(ay) - x^2(ya) \\
&\quad + x[(xa)y + (xy)a] \\
&= x[-x(ay) - x(ya) + (xa)y + (xy)a] \\
&= x[(x, a, y) + (x, y, a)] = 0,
\end{aligned}
$$

establishing (3.4). Identity (3.5) is the reciprocal relationship (obtained by passing to the anti-isomorphic algebra, which is alternative since the defining identities are reciprocal). Finally (3.4) implies

$$
\begin{aligned}
(xy)(ax) - x(ya)x &= (x, y, ax) + x[y(ax) - (ya)x] \\
&= -(x, ax, y) - x(y, a, x) \\
&= -(xax)y + x[(ax)y - (y, a, x)] \\
&= -x[a(xy) - (ax)y + (y, a, x)] \\
&= -x[-(a, x, y) + (y, a, x)] = 0.
\end{aligned}
$$

The Moufang identity (3.5) is equivalent to

$$(3.7) \qquad (y, xa, x) = -(y, x, a)x$$

for all x, y, a in $\mathfrak{A}$, since

$$(y, xa, x) = [y(xa)]x - y(xax) = [y(xa)]x - [(yx)a]x = -(y, x, a)x.$$

The linearized form of (3.7) is

$$(3.8) \qquad (y, xa, z) + (y, za, x) = -(y, x, a)z - (y, z, a)x$$

for all x, y, z, a in $\mathfrak{A}$.

Theorem 3.1 (Artin). *The subalgebra generated by any two elements of an alternative algebra $\mathfrak{A}$ is associative.*

Proof. If x, y are any two elements of $\mathfrak{A}$, we denote by $p = p(x, y)$ any nonassociative product $z_1 z_2 \cdots z_t$ (with some distribution of parentheses) of t factors z_i, each of which is equal to either x or y. Also we denote the degree t of such a product by ∂p. It is sufficient for us to prove that $(p, q, r) = 0$ for all nonassociative products $p = p(x, y)$, $q = q(x, y)$, $r = r(x, y)$. We shall prove this by induction on $n = \partial p + \partial q + \partial r$. The result is vacuous for $n < 3$, and we assume $(p, q, r) = 0$ whenever $\partial p + \partial q + \partial r < n$. Since $\partial p < n$, the induction hypothesis implies, by the usual argument which yields the generalized associative law from the associative law, that parentheses are not necessary in the product $p = p(x, y) = z_1 z_2 \cdots z_t$, and we say that p begins with z_1.

Now two of the products p, q, r must begin with the same letter, say x. Since associators alternate in $\mathfrak{A}$, we may assume that q and r begin with x.

If $\partial q > 1$ and $\partial r > 1$, then $q = xq'$, $r = xr'$ where $\partial q' = -1 + \partial q$, $\partial r' = -1 + \partial r$. Putting $y = xr'$, $a = q'$, $z = p$ in (3.8), we have

$$(p, q, r) = (p, xq', xr') = -(xr', xq', p)$$
$$= (xr', pq', x) + (xr', x, q')p + (xr', p, q')x$$
$$= -(pq', xr', x) = (pq', x, r')x = 0$$

by (3.7) and the assumption of the induction. If only one of the products q, r has degree > 1 (say $q = xq'$), then (3.7) implies that

$$(p, q, r) = (p, xq', x) = -(p, x, q')x = 0$$

by the assumption of the induction. The easiest case $\partial q = \partial r = 1$ is given by the right alternative law:

$$(p, q, r) = (p, x, x) = 0.$$

For more general theorems on associativity in alternative algebras, see Bruck and Kleinfeld [1].

An algebra $\mathfrak{A}$ over F is called *power-associative* in case the subalgebra $F[x]$ of $\mathfrak{A}$ generated by any element x in $\mathfrak{A}$ is associative. This is equivalent to defining powers of a single element x in $\mathfrak{A}$ recursively by $x^1 = x$, $x^{i+1} = xx^i$, and requiring that

(3.9) $x^i x^j = x^{i+j}$ for all x in $\mathfrak{A}$ $(i, j = 1, 2, 3, \ldots)$.

Theorem 3.1 implies that any alternative algebra is power-associative. We could have proved this directly by induction on i in (3.9). Theorem 3.1 also implies

(3.10) $R_x{}^j = R_{x^j}$, $L_x{}^j = L_{x^j}$ for all x in $\mathfrak{A}$.

An element x in a power-associative algebra $\mathfrak{A}$ is called *nilpotent* in case there is an integer r such that $x^r = 0$. An algebra (ideal) consisting only of nilpotent elements is called a *nilalgebra* (*nilideal*).

Theorem 3.2.   *Any alternative nilalgebra $\mathfrak{A}$ of finite dimension over F is nilpotent.*

Proof.   Let $\mathfrak{B}$ be a subalgebra of $\mathfrak{A}$ which is maximal with respect to the property that $\mathfrak{B}^*$ is nilpotent. Such a maximal $\mathfrak{B}$ exists, since the subalgebra $\{0\}$ has this property and $\mathfrak{A}$ is finite-dimensional.

Assume that $\mathfrak{B}$ is a proper subalgebra of $\mathfrak{A}$. Then there exists an element x not in $\mathfrak{B}$ such that

$$(3.11) \qquad\qquad x\mathfrak{B}^* \subseteq \mathfrak{B}.$$

For $\mathfrak{B}^{*r} = 0$ implies that $\mathfrak{A}\mathfrak{B}^{*r} = 0 \subseteq \mathfrak{B}$, and there exists a smallest integer $m \geq 1$ such that $\mathfrak{A}\mathfrak{B}^{*m} \subseteq \mathfrak{B}$. If $m = 1$, take x in $\mathfrak{A}$ but not in $\mathfrak{B}$; if $m > 1$, take x in $\mathfrak{A}\mathfrak{B}^{*m-1}$ but not in $\mathfrak{B}$. Then (3.11) is satisfied. Let

$$(3.12) \qquad\qquad \mathfrak{C} = \mathfrak{B} + F[x].$$

Then $\mathfrak{C}^* = (\mathfrak{B} + Fx)^*$. Put $y = b$ in (3.2) for any b in $\mathfrak{B}$. Then (3.11) implies that

$$(3.13) \quad \begin{aligned} R_x R_b &= R_{b_1} - R_b R_x, & R_x L_b &= L_b R_x + R_b R_x - R_{b_2}, \\ L_x R_b &= R_b L_x + L_b L_x - L_{b_3}, & L_x L_b &= L_{b_1} - L_b L_x \end{aligned}$$

for b_i in $\mathfrak{B}$. Equations (3.13) show that, in each product of right and left multiplications in $\mathfrak{B}^*$ and $(Fx)^*$, the multiplication R_x or L_x may be systematically passed from the left to the right of R_b or L_b in a fashion which, although it may change signs and introduce new terms, preserves the number of factors from $\mathfrak{B}^*$, and does not increase the number of factors from $(Fx)^*$. Hence any T in $\mathfrak{C}^* = (\mathfrak{B} + Fx)^*$ may be written as a linear combination of terms of the form

$$R_x^{j_1}, \quad L_x^{j_2}, \quad R_x^{j_3}L_x^{j_4}, \quad B_1, \quad B_2 R_x^{m_1}, \quad B_3 L_x^{m_2}, \quad B_4 R_x^{m_3}L_x^{m_4}$$

for B_i in $\mathfrak{B}^*$, $j_i \geq 1$, $m_i \geq 1$. Then, if $\mathfrak{B}^{*r} = 0$ and $x^j = 0$, we have $T^{r(2j-1)} = 0$; for every term in the expansion of $T^{r(2j-1)}$ contains either an uninterrupted sequence of at least $2j - 1$ factors from $(Fx)^*$ or at least r factors B_i. In the first case the term contains either a product $R_x{}^j = R_{x^j} = 0$ or a product $L_x{}^j = L_{x^j} = 0$ by (3.10). In the latter case the R_x or L_x may be systematically passed from the left to the right of B_i (as above) preserving the number of factors from $\mathfrak{B}^*$, resulting in a sum of terms each containing a product $B_1 B_2 \cdots B_r = 0$. Hence every element T of the finite-dimensional associative algebra $\mathfrak{C}^*$ is nilpotent. Hence the associative algebra $\mathfrak{C}^*$ is nilpotent (Albert [24], p. 23). But since $\mathfrak{B}$ is maximal with respect to the property of having $\mathfrak{B}^*$ nilpotent, this is a contradiction. Hence $\mathfrak{B}$ is not a proper subalgebra of $\mathfrak{A}$. That is, $\mathfrak{B} = \mathfrak{A}$ and $\mathfrak{A}^*$ is nilpotent. Hence $\mathfrak{A}$ is nilpotent by Theorem 2.4.

Any nilpotent algebra is solvable, and any solvable (power-associative) algebra is a nilalgebra. By Theorem 3.2 the concepts of nilpotent algebra, solvable algebra, and nilalgebra coincide for finite-dimensional alternative algebras. Hence there is a unique maximal nilpotent ideal $\mathfrak{N}$ (= solvable ideal = nilideal) in any finite-dimensional alternative algebra $\mathfrak{A}$; we call $\mathfrak{N}$ the *radical* of $\mathfrak{A}$. We have seen that the radical of $\mathfrak{A}/\mathfrak{N}$ is 0.

2.   THE PEIRCE DECOMPOSITION

We say that $\mathfrak{A}$ is *semisimple* in case the radical of $\mathfrak{A}$ is 0, and prove in Section 3 that any finite-dimensional semisimple alternative algebra $\mathfrak{A}$ is the direct sum $\mathfrak{A} = \mathfrak{S}_1 \oplus \cdots \oplus \mathfrak{S}_t$ of simple algebras $\mathfrak{S}_i$. The proof is dependent upon the properties of the *Peirce decomposition* relative to a set of pairwise orthogonal idempotents $e_1, \ldots, e_t$ in $\mathfrak{A}$.

An element e of an (arbitrary) algebra $\mathfrak{A}$ is called an *idempotent* in case $e^2 = e \neq 0$.

Proposition 3.3.   *Any finite-dimensional power-associative algebra $\mathfrak{A}$, which is not a nilalgebra, contains an idempotent e ($\neq 0$).*

Proof. $\mathfrak{A}$ contains an element x which is not nilpotent. The subalgebra $F[x]$ of $\mathfrak{A}$ generated by x is a finite-dimensional associative algebra which is not a nilalgebra. Then $F[x]$ contains an idempotent e ($\neq 0$) (Albert [24], p. 23), and therefore $\mathfrak{A}$ does.

By (3.1) L_e and R_e are idempotent operators on $\mathfrak{A}$ which commute by the flexible law (they are "commuting projections"). It follows that $\mathfrak{A}$ is the vector space direct sum

$$(3.14) \qquad \mathfrak{A} = \mathfrak{A}_{11} + \mathfrak{A}_{10} + \mathfrak{A}_{01} + \mathfrak{A}_{00},$$

where $\mathfrak{A}_{ij}$ $(i, j = 0, 1)$ is the subspace of $\mathfrak{A}$ defined by

$$\mathfrak{A}_{ij} = \{x_{ij} \mid ex_{ij} = ix_{ij}, \ x_{ij}e = jx_{ij}\}, \qquad i, j = 0, 1.$$

Just as in the case of associative algebras, the decomposition of any element x in $\mathfrak{A}$ according to the *Peirce decomposition* (3.14) is

$$x = exe + (ex - exe) + (xe - exe) + (x - ex - xe + exe).$$

We derive a few of the properties of the Peirce decomposition relative to a single idempotent e as follows. For $i, j, k, l = 0, 1$, we have

$$e(x_{ij} y_{kl}) = -(e, x_{ij}, y_{kl}) + (ex_{ij})y_{kl}$$
$$= (x_{ij}, e, y_{kl}) + (ex_{ij})y_{kl}$$
$$= jx_{ij}y_{kl} - kx_{ij}y_{kl} + ix_{ij}y_{kl}$$
$$= (i + j - k)x_{ij}y_{kl},$$

and similarly $(x_{ij} y_{kl})e = (k + l - j)x_{ij}y_{kl}$, so

$$\mathfrak{A}_{ij}\mathfrak{A}_{jk} \subseteq \mathfrak{A}_{ik}, \qquad i, j, k = 0, 1;$$

and

$$\mathfrak{A}_{ij}\mathfrak{A}_{ij} \subseteq \mathfrak{A}_{ji}, \qquad i, j = 0, 1.$$

In particular, $\mathfrak{A}_{11}$ and $\mathfrak{A}_{00}$ are subalgebras of $\mathfrak{A}$. Also $x_{11}y_{00} = (ex_{11}e)y_{00} = e[x_{11}(ey_{00})] = 0$ by (3.4), and similarly $y_{00}x_{11} = 0$. Hence $\mathfrak{A}_{11}$ and $\mathfrak{A}_{00}$ are orthogonal subalgebras of $\mathfrak{A}$. Also $\mathfrak{A}_{10}\mathfrak{A}_{01} \subseteq \mathfrak{A}_{11}$, $\mathfrak{A}_{01}\mathfrak{A}_{10} \subseteq \mathfrak{A}_{00}$, etc. The properties $\mathfrak{A}_{10}\mathfrak{A}_{10} \subseteq \mathfrak{A}_{01}$, $\mathfrak{A}_{01}\mathfrak{A}_{01} \subseteq \mathfrak{A}_{10}$ are weaker than in the associative case, where one can actually prove that $\mathfrak{A}_{10}\mathfrak{A}_{10} = \mathfrak{A}_{01}\mathfrak{A}_{01} = 0$.

Idempotents $e_1, \ldots, e_t$ in an (arbitrary) algebra $\mathfrak{A}$ are called *pairwise orthogonal* in case $e_i e_j = 0$ for $i \neq j$. Note that any sum $e = e_1 + \cdots + e_t$ of pairwise orthogonal idempotents ($t \geq 1$) is an idempotent. Also $ee_i = e_i e = e_i$ ($i = 1, \ldots, t$).

In an alternative algebra each associator $(x, e_i, e_j) = 0$. [If $i = j$, the right alternative law implies this; if $i \neq j$, then

$$(x, e_i, e_j) = (xe_i)e_j = (xe_i)e_j^2 = (x, e_i, e_j)e_j$$
$$= -(x, e_j, e_i)e_j = -((xe_j)e_i)e_j = -x(e_j e_i e_j) = 0$$

by (3.5).] Also $(x, e_i, e) = 0$ for e above.

A more refined Peirce decomposition for an alternative algebra $\mathfrak{A}$ than the one given above is the following decomposition relative to a set $e_1, \ldots, e_t$ of pairwise orthogonal idempotents in $\mathfrak{A}$: $\mathfrak{A}$ is the vector space direct sum

(3.15) $$\mathfrak{A} = \sum \mathfrak{A}_{ij} \qquad (i, j = 0, 1, \ldots, t),$$

where, using the Kronecker delta, we define

(3.16) $\quad \mathfrak{A}_{ij} = \{x_{ij} \mid e_k x_{ij} = \delta_{ki} x_{ij}, \quad x_{ij} e_k = \delta_{jk} x_{ij} \text{ for } \quad k = 1, \ldots, t\}$

$(i, j = 0, 1, \ldots, t)$. The $\mathfrak{A}_{ij}$ are clearly subspaces of $\mathfrak{A}$. We prove first the uniqueness of an expression

$$x = \sum_{k,l=0}^{t} x_{kl}, \qquad x_{kl} \in \mathfrak{A}_{kl}.$$

For $i, j = 1, \ldots, t$, we have

$$e_i x e_j = \sum_{k,l=0}^{t} e_i x_{kl} e_j = x_{ij}.$$

For $i = 1, \ldots, t$, we have

$$e_i x = \sum_{k,l=0}^{t} e_i x_{kl} = \sum_{l=0}^{t} x_{il}.$$

Then

$$e_i x e = \sum_{l=0}^{t} x_{il} \sum_{k=1}^{t} e_k = \sum_{l=1}^{t} x_{il},$$

so $e_i x - e_i x e = x_{i0}$, where $e = e_1 + \cdots + e_t$. Similarly, $xe_j - exe_j = x_{0j}$ for $j = 1, \ldots, t$. Finally,

$$x_{00} = x - \sum_{i,j=1}^{t} x_{ij} - \sum_{i=1}^{t} x_{i0} - \sum_{j=1}^{t} x_{0j}$$

$$= x - \sum_{i,j=1}^{t} e_i x e_j - \sum_{i=1}^{t} (e_i x - e_i x e) - \sum_{j=1}^{t} (x e_j - e x e_j)$$

$$= x - exe - (ex - exe) - (xe - exe) = x - ex - xe + exe.$$

But these x_{ij} are in $\mathfrak{A}_{ij}$ $(i, j = 0, 1, \ldots, t)$, so x in $\mathfrak{A}$ is uniquely expressible in the form $x = \sum x_{ij}$, x_{ij} in $\mathfrak{A}_{ij}$, and $\mathfrak{A}$ is the vector space direct sum (3.15). The expression

$$x = \sum_{i,j=0}^{t} x_{ij}$$

$$= \sum_{i,j=1}^{t} e_i x e_j + \sum_{i=1}^{t} (e_i x - e_i x e) + \sum_{j=1}^{t} (x e_j - e x e_j)$$

$$+ (x - ex - xe + exe)$$

is a refinement of the decomposition of x relative to a single idempotent e.

Proposition 3.4 (Properties of the Peirce decomposition for alternative algebras).

(3.17) $\quad\quad\quad \mathfrak{A}_{ij}\mathfrak{A}_{jk} \subseteq \mathfrak{A}_{ik} \quad\quad\quad\quad (i, j, k = 0, 1, \dots, t);$

(3.18) $\quad\quad\quad \mathfrak{A}_{ij}\mathfrak{A}_{ij} \subseteq \mathfrak{A}_{ji} \quad\quad\quad\quad (i, j = 0, 1, \dots, t);$

(3.19) $\quad\quad\quad \mathfrak{A}_{ij}\mathfrak{A}_{kl} = 0, \quad j \neq k, \quad (i, j) \neq (k, l)$

$$(i, j, k, l = 0, 1, \dots, t);$$

(3.20) $\quad\quad\quad x_{ij}^2 = 0 \quad\quad\quad \text{for all} \quad x_{ij} \text{ in } \mathfrak{A}_{ij} \quad (i \neq j),$

implying

(3.21) $\quad\quad\quad x_{ij}y_{ij} = -y_{ij}x_{ij} \quad \text{for all} \quad x_{ij}, y_{ij} \text{ in } \mathfrak{A}_{ij} \quad (i \neq j);$

(3.22) $\quad (x_{ij}, y_{jk}, z_{ki}) = 0 \quad \text{if} \quad (i, j, k) \neq (i, i, i), \quad \text{for all} \quad x_{ij}$

$$\text{in } \mathfrak{A}_{ij}, \quad y_{jk} \text{ in } \mathfrak{A}_{jk}, \quad z_{ki} \text{ in } \mathfrak{A}_{ki};$$

(3.23) $\quad (x_{ii}, y_{ij}z_{ji}, t_{ii}) = 0 \quad \text{if} \quad i \neq j \quad \text{for all} \quad x_{ii}, t_{ii} \text{ in } \mathfrak{A}_{ii}, y_{ij}$

$$\text{in } \mathfrak{A}_{ij}, z_{ji} \text{ in } \mathfrak{A}_{ji};$$

(3.24) $\quad\quad (x_{ij}y_{ij})z_{ij} = (y_{ij}z_{ij})x_{ij} = (z_{ij}x_{ij})y_{ij} \quad \text{if} \quad i \neq j$

$$\text{for all} \quad x_{ij}, y_{ij}, z_{ij} \text{ in } \mathfrak{A}_{ij};$$

(3.25) $\quad x_{ij}(y_{ij}z_{jj}) = (x_{ij}z_{jj})y_{ij} = z_{jj}(x_{ij}y_{ij}) \quad \text{if} \quad i \neq j$

$$\text{for all} \quad x_{ij}, y_{ij} \text{ in } \mathfrak{A}_{ij}; z_{jj} \text{ in } \mathfrak{A}_{jj};$$

and, reciprocally,

(3.26) $\quad x_{ij}(z_{ii}y_{ij}) = (z_{ii}x_{ij})y_{ij} = (x_{ij}y_{ij})z_{ii} \quad \text{if} \quad i \neq j$

$$\text{for all} \quad x_{ij}, y_{ij} \text{ in } \mathfrak{A}_{ij}; z_{ii} \text{ in } \mathfrak{A}_{ii};$$

(3.27) $\quad [x_{jj}, t_{jj}](y_{ij}z_{ij}) = 0 \quad \text{if} \quad i \neq j \quad \text{for all} \quad x_{jj}, t_{jj} \text{ in } \mathfrak{A}_{jj};$

$$y_{ij}, z_{ij} \text{ in } \mathfrak{A}_{ij};$$

and, reciprocally,

(3.28) $(y_{ij} z_{ij})[x_{ii}, t_{ii}] = 0$ *if* $i \neq j$ *for all* x_{ii}, t_{ii} *in* $\mathfrak{A}_{ii}$; y_{ij}, z_{ij} *in* $\mathfrak{A}_{ij}$;

(3.29) $(x_{ii} a)^m = (x_{ii} a_{ii})^{m-1} \sum\limits_{k=0}^{t} x_{ii} a_{ik}$;

(3.30) $(x_{ij} a)^m = (x_{ij} a_{ji})^{m-1} \sum\limits_{k=0}^{t} x_{ij} a_{jk} + (x_{ij} a_{ij})(x_{ij} a_{ji})^{m-1}$

            if $i \neq j$;

(3.31) $e_i(x_{ij} a)^m e_i = (x_{ij} a_{ji})^m$ $(i, j = 0, 1, \dots, t)$.

Proof of (3.17) *and* (3.18).   We have

$$e_h(x_{ij} y_{kl}) = (e_h x_{ij}) y_{kl} - (e_h, x_{ij}, y_{kl})$$
$$= \delta_{hi} x_{ij} y_{kl} + (x_{ij}, e_h, y_{kl})$$
$$= (\delta_{hi} + \delta_{jh} - \delta_{hk}) x_{ij} y_{kl};$$
$$(x_{ij} y_{kl}) e_h = x_{ij}(y_{kl} e_h) + (x_{ij}, y_{kl}, e_h)$$
$$= \delta_{lh} x_{ij} y_{kl} - (x_{ij}, e_h, y_{kl})$$
$$= (\delta_{lh} - \delta_{jh} + \delta_{hk}) x_{ij} y_{kl}.$$

Proof of (3.19) *if* $k \neq 0 \neq l$, $k \neq l \neq j$. When the associators in (3.8) are expanded, we obtain the linearized form of (3.5):

$$y[(xa)z] + y[(za)x] = [(yx)a]z + [(yz)a]x.$$

Putting $y = x_{ij}$, $x = e_k$, $a = y_{kl}$, $z = e_l$ in this, we have

$$x_{ij}(e_k y_{kl} e_l) + x_{ij}(e_l y_{kl} e_k) = [(x_{ij} e_k) y_{kl}] e_l + [(x_{ij} e_l) y_{kl}] e_k,$$

so that $x_{ij} y_{kl} = 0$.

Proof of (3.20).   By (3.18) x_{ij}^2 is in $\mathfrak{A}_{ji}$. Either $i \neq 0$ or $j \neq 0$. If $i \neq 0$, then

$$x_{ij}^2 = x_{ij}^2 e_i = x_{ij}(x_{ij} e_i) = 0,$$

since $i \neq j$. Reciprocally, if $j \neq 0$, then $x_{ij}^2 = 0$.

Proof of (3.23). We have

$$(x_{ii}, y_{ij}z_{ji}, t_{ii}) = ((x_{ii}y_{ij})z_{ji})t_{ii} - (x_{ii}, y_{ij}, z_{ji})t_{ii}$$
$$- x_{ii}(y_{ij}(z_{ji}t_{ii})) - x_{ii}(y_{ij}, z_{ji}, t_{ii})$$
$$= (x_{ii}y_{ij}, z_{ji}, t_{ii}) + (x_{ii}, y_{ij}, z_{ji}t_{ii}) = 0$$

by (3.17) and (3.22).

Proof of (3.29) *by induction on m.* By (3.19) we have

$$x_{ii}a = x_{ii}\sum_{j,k} a_{jk} = \sum_k x_{ii}a_{ik},$$

giving the case $m = 1$. Assume (3.29). Then

$$(x_{ii}a)^{m+1} = (x_{ii}a)^m(x_{ii}a) = (x_{ii}a_{ii})^{m-1}\left(\sum_{k=0}^{t}(x_{ii}a_{ik})(x_{ii}a)\right)$$
$$= (x_{ii}a_{ii})^{m-1}\sum_{j,k=0}^{t}(x_{ii}a_{ik})(x_{ii}a_{ij})$$
$$= (x_{ii}a_{ii})^{m-1}\sum_{j=0}^{t}(x_{ii}a_{ii})(x_{ii}a_{ij}) = (x_{ii}a_{ii})^m\sum_{k=0}^{t}(x_{ii}a_{ik}),$$

as desired, by (3.17), (3.19), and (3.20).

The proofs of the remaining cases of (3.19) and of (3.22), (3.24), (3.25), (3.27), (3.30), and (3.31) are left as exercises for the reader.

3. THE RADICAL; SEMISIMPLE ALGEBRAS

An element z of an alternative algebra $\mathfrak{A}$ is called *properly nilpotent* (in $\mathfrak{A}$) in case za is nilpotent for all a in $\mathfrak{A}$. Note that z is properly nilpotent if and only if az is nilpotent for all a in $\mathfrak{A}$. For $(az)^{m+1} = a(za)^m z$, etc., by Theorem 3.1. Also z properly nilpotent implies z is nilpotent (since z^2 is).

We shall show that the radical ($=$ maximal nilideal) $\mathfrak{N}$ of any finite-dimensional alternative algebra $\mathfrak{A}$ is the set $\mathfrak{P}$ of all properly nilpotent elements of $\mathfrak{A}$. Clearly $\mathfrak{N} \subseteq \mathfrak{P}$, since z in $\mathfrak{N}$ implies za is in $\mathfrak{N}$ for all a in $\mathfrak{A}$, whence za is nilpotent for all a in $\mathfrak{A}$. In the associative case, the proof that $\mathfrak{P} \subseteq \mathfrak{N}$ requires only a few lines (Albert [24], p. 24). The proof, due to Zorn, of the generalization to alternative algebras employs most of the properties of the Peirce decomposition listed above.

In any algebra $\mathfrak{A}$ with 1, an element x is said to have an *inverse* x^{-1} in case there is an x^{-1} in $\mathfrak{A}$ satisfying $xx^{-1} = x^{-1}x = 1$. In an alternative algebra, if x has an inverse, it is unique. Also $(x, x^{-1}, y) = 0$. For we may put $a = x^{-1}$, $y = x$, $z = x^{-1}y$ in (2.4) to obtain $(x^{-1}, x^2, x^{-1}y) + (x, x^{-1}, x(x^{-1}y)) = 0$. Then the Moufang identity (3.4) implies that $(x, x^{-1}, y) = 0$. Suppose that x has another (right) inverse x': $xx' = 1$. Then

$$x' = 1x' = (x^{-1}x)x' = x^{-1}(xx') = x^{-1}1 = x^{-1}.$$

If every nonzero element in an alternative algebra $\mathfrak{A}$ with 1 has an inverse, then $\mathfrak{A}$ is a division algebra, and conversely. For the equations (2.5) have unique solutions $x = a^{-1}b$, $y = ba^{-1}$.

Lemma 3.5. *Let $\mathfrak{A}$ be a finite-dimensional alternative algebra over F with 1, and let 1 be the only idempotent in $\mathfrak{A}$. Then every element z of $\mathfrak{A}$ either has an inverse z^{-1} in $\mathfrak{A}$ or is properly nilpotent. The set $\mathfrak{P}$ of properly nilpotent elements of $\mathfrak{A}$ is an ideal of $\mathfrak{A}$.*

Proof. If z is not nilpotent, then z has an inverse in $\mathfrak{A}$. For $F[z]$ contains an idempotent $(= 1)$:

$$1 = \alpha_0 z^n + \cdots + \alpha_{n-1}z, \qquad \alpha_i \text{ in } F,$$

so

$$1 = z(\alpha_0 z^{n-1} + \cdots + \alpha_{n-1}1) = zy, \qquad y \text{ in } F1 + F[z] \subseteq \mathfrak{A}.$$

Then $yz = 1$, so $y = z^{-1}$. But any nilpotent element z of $\mathfrak{A}$ is properly nilpotent. For a in $\mathfrak{A}$ implies that za is nilpotent; if not, za has an inverse $(za)^{-1}$ in $\mathfrak{A}$. Let $z^m = 0$, $z^{m-1} \neq 0$. Then

$$0 \neq z^{m-1} = z^{m-1}[(za)(za)^{-1}] = [z^{m-1}(za)](za)^{-1} = (z^m a)(za)^{-1} = 0,$$

a contradiction. Let z be in $\mathfrak{P}$. Then za is nilpotent (and therefore properly nilpotent) for every a in $\mathfrak{A}$. Hence za is in $\mathfrak{P}$. Similarly az is in $\mathfrak{P}$ for all a in $\mathfrak{A}$. It remains to show that $\mathfrak{P}$ is a subspace of $\mathfrak{A}$. Clearly αz is in $\mathfrak{P}$ for all z in $\mathfrak{P}$, α in F. Let z, z' be in $\mathfrak{P}$. If $z + z'$ is not in $\mathfrak{P}$, then $z + z'$ has an inverse y: $(z + z')y = 1$. Then $z'y = 1 - zy$ has an inverse, since zy is nilpotent $[(1 - zy)^{-1} = 1 + zy + (zy)^2 + \cdots]$. But z' in $\mathfrak{P}$ implies that $z'y$ is nilpotent, a contradiction. Hence $\mathfrak{P}$ is an ideal of $\mathfrak{A}$.

An idempotent e in an (arbitrary) algebra $\mathfrak{A}$ is called *primitive* in case there do not exist orthogonal idempotents u, v in $\mathfrak{A}$ ($uv = vu = 0$) such that $e = u + v$. In a finite-dimensional algebra $\mathfrak{A}$, any idempotent e may be written as a sum $e = e_1 + \cdots + e_t$ of pairwise orthogonal primitive idempotents e_i ($t = 1$ in case e is primitive). For if e is not primitive, we have $e = u + v$ and, continuing if either u or v is not primitive, etc., we have $e = u_1 + \cdots + u_r$ where the pairwise orthogonal idempotents $u_1, \ldots, u_r$ span an r-dimensional subspace (commutative associative subalgebra) of $\mathfrak{A}$. Since $r \leq \dim \mathfrak{A}$, the process terminates with pairwise orthogonal primitive idempotents e_i, $e = e_1 + \cdots + e_t$.

In any algebra $\mathfrak{A}$ over F an idempotent e is called *principal* in case there is no idempotent u in $\mathfrak{A}$ which is orthogonal to e ($u^2 = u \neq 0$, $ue = eu = 0$). In a finite-dimensional alternative algebra $\mathfrak{A}$, this means that e is a principal idempotent of $\mathfrak{A}$ if and only if the subalgebra $\mathfrak{A}_{00}$, in the Peirce decomposition (3.14) relative to e, is a nilalgebra.

Now any finite-dimensional alternative algebra $\mathfrak{A}$ which is not a nilalgebra contains a principal idempotent. For $\mathfrak{A}$ contains an idempotent e by Proposition 3.3. If e is not principal, there is an idempotent u in $\mathfrak{A}_{00}$, $e' = e + u$ is idempotent, and $\mathfrak{A}_{11,e'}$ (the $\mathfrak{A}_{11}$ relative to e') contains properly $\mathfrak{A}_{11,e} = \mathfrak{A}_{11}$. For x_{11} in $\mathfrak{A}_{11,e}$ implies

$$x_{11}e' = x_{11}(e + u) = x_{11}e + x_{11}u = x_{11},$$

and similarly $e'x_{11} = x_{11}$, so that x_{11} is in $\mathfrak{A}_{11,e'}$. That is, $\mathfrak{A}_{11,e} \subseteq \mathfrak{A}_{11,e'}$. But $u \in \mathfrak{A}_{11,e'}$, $u \notin \mathfrak{A}_{11,e}$. Then $\dim \mathfrak{A}_{11,e} < \dim \mathfrak{A}_{11,e'}$, and this process of increasing dimensions must terminate, yielding a principal idempotent.

We note that, if u is any idempotent in a finite-dimensional alternative algebra $\mathfrak{A}$, then there exist pairwise orthogonal primitive idempotents $e_1, \ldots, e_r, \ldots, e_t$ in $\mathfrak{A}$ ($1 \leq r \leq t$) such that $u = e_1 + \cdots + e_r$ while $e = e_1 + \cdots + e_t$ is a principal idempotent in $\mathfrak{A}$.

In order to see that the set $\mathfrak{P}$ of properly nilpotent elements in a finite-dimensional alternative algebra $\mathfrak{A}$ is the radical $\mathfrak{N}$ of $\mathfrak{A}$, we use the Peirce decomposition (3.15) of $\mathfrak{A}$ relative to a set $e_1, \ldots, e_t$ of pairwise orthogonal primitive idempotents e_i such that $e = e_1 + \cdots + e_t$ is a principal idempotent of $\mathfrak{A}$. Then $\mathfrak{A}_{00}$ is a nilalgebra, and e_i is the unity element for $\mathfrak{A}_{ii}$ ($i = 1, \ldots, t$). Moreover, e_i is the only idempotent in $\mathfrak{A}_{ii}$ ($i = 1, \ldots, t$). For, if u in $\mathfrak{A}_{ii}$ is idempotent, $u \neq e_i$, then $e_i = u + (e_i - u)$ implies that e_i is not primitive. Hence Lemma 3.5 may be applied to each $\mathfrak{A}_{ii}$ ($i = 1, \ldots, t$). By (3.29) any element x_{ii} in $\mathfrak{A}_{ii}$ which is properly nilpotent in $\mathfrak{A}_{ii}$ is properly nilpotent in $\mathfrak{A}$ $(i = 0, 1, \ldots, t)$. For $i, j = 0, 1, \ldots, t$, let

$$\mathfrak{S}_{ij} = \{s_{ij} \in \mathfrak{A}_{ij} \mid \text{all elements of } s_{ij}\mathfrak{A}_{ji} \text{ are nilpotent}\},$$

and note that

$$\mathfrak{S}_{ij} = \{s_{ij} \in \mathfrak{A}_{ij} \mid \text{all elements of } \mathfrak{A}_{ji}s_{ij} \text{ are nilpotent}\}.$$

Lemma 3.6. *With the notation above, each $\mathfrak{S}_{ij}$ $(i, j = 0, 1, \ldots, t)$ is a subspace of $\mathfrak{A}$, and $\mathfrak{S}_{ij} \subseteq \mathfrak{P}$, the set of properly nilpotent elements of $\mathfrak{A}$.*

Proof. Since $\mathfrak{A}_{00}$ is a nilalgebra, we have seen above that $\mathfrak{S}_{00} = \mathfrak{A}_{00} \subseteq \mathfrak{P}$. Also we have $\mathfrak{S}_{0j} = \mathfrak{A}_{0j}$ $(j = 1, \ldots, t)$, and $\mathfrak{A}_{0j} \subseteq \mathfrak{P}$ by (3.30). To see that $\mathfrak{S}_{i0} = \mathfrak{A}_{i0} \subseteq \mathfrak{P}$ $(i = 1, \ldots, t)$, we use

$$(x_{i0}a_{0i})^{m-1} = x_{i0}(a_{0i}x_{i0})^{m-2}a_{0i}$$

in (3.30). Thus we need only consider $\mathfrak{S}_{ij}$ with $i \neq 0$, $j \neq 0$. Suppose that s_{ij}, s'_{ij} are in $\mathfrak{S}_{ij}$; then $s_{ij}a_{ji}$ and $s'_{ij}a_{ji}$ in $\mathfrak{A}_{ii}$ belong to the ideal of nilpotent elements of $\mathfrak{A}_{ii}$, and $(\alpha s_{ij} + \beta s'_{ij})a_{ji}$ is nilpotent for all α, β in F. That is, $\alpha s_{ij} + \beta s'_{ij}$ is in $\mathfrak{S}_{ij}$, and $\mathfrak{S}_{ij}$ is a subspace of $\mathfrak{A}$. Finally, (3.29) and (3.30) imply that each s_{ij} in $\mathfrak{S}_{ij}$ $(i, j = 1, \ldots, t)$ is properly nilpotent.

Theorem 3.7 (Zorn). *The radical $\mathfrak{R}$ $(= $ maximal nilideal) of any finite-dimensional alternative algebra $\mathfrak{A}$ is the set $\mathfrak{P}$ of all properly nilpotent elements of $\mathfrak{A}$.*

Proof. With the notation above, let $\mathfrak{S} = \sum \mathfrak{S}_{ij}$. Also denote by $\tilde{\mathfrak{P}}$ the subspace of $\mathfrak{A}$ spanned by $\mathfrak{P}$. Then $\mathfrak{S} \subseteq \tilde{\mathfrak{P}}$ by Lemma 3.6. We shall show that $\mathfrak{S} = \tilde{\mathfrak{P}}$ is an ideal of $\mathfrak{A}$.

Suppose there exists $x \in \mathfrak{P}$, $x \notin \mathfrak{S} = \sum \mathfrak{S}_{ij}$. Then $x = \sum x_{ij}$, $x_{ij} \in \mathfrak{A}_{ij}$, with at least one $x_{ij} \notin \mathfrak{S}_{ij}$. As indicated in the proof of Lemma 3.6, we then have $i \neq 0$, $j \neq 0$ for this particular element x_{ij}. If $i = j$, then by Lemma 3.5 the element $x_{ij} = x_{jj}e_j \in \mathfrak{A}_{jj}$, being not nilpotent, has an inverse b_{ji} with respect to e_j: $b_{ji}x_{ij} = e_j$. If $i \neq j$, we have the same equation

$$(3.32) \qquad b_{ji}x_{ij} = e_j \qquad \text{for some} \quad b_{ji} \text{ in } \mathfrak{A}_{ji}.$$

For there exists a_{ji} in $\mathfrak{A}_{ji}$ such that $x_{ij}a_{ji}$ is not nilpotent. But then $a_{ji}x_{ij}$ in $\mathfrak{A}_{jj}$ is not nilpotent and has an inverse z_{jj} with respect to e_j: $z_{jj}(a_{ji}x_{ij})$

$= e_j$. By (3.22) we have (3.32) where in this case $(i \neq j)$ $b_{ji} = z_{jj} a_{ji}$. That is, (3.32) holds for both $i = j$ and $i \neq j$. Replacing i by j, j by i, x_{ij} by b_{ji}, a by x in (3.31), we have

$$e_j(b_{ji}x)^m e_j = (b_{ji}x_{ij})^m = e_j{}^m = e_j \neq 0,$$

so that $b_{ji}x$ is not nilpotent, $x \notin \mathfrak{P}$, a contradiction. Hence $\mathfrak{P} \subseteq \mathfrak{S}$. Since $\mathfrak{S}$ is a subspace, $\tilde{\mathfrak{P}} \subseteq \mathfrak{S}$; hence $\mathfrak{S} = \tilde{\mathfrak{P}}$.

To see that $\mathfrak{S}$ is a right ideal of $\mathfrak{A}$, we require

(3.33) $$\mathfrak{S}_{ij}\mathfrak{A}_{jk} \subseteq \mathfrak{S}_{ik} \qquad (i, j, k = 0, 1, \ldots, t)$$

and

(3.34) $$\mathfrak{S}_{ij}\mathfrak{A}_{ij} \subseteq \mathfrak{S}_{ji} \qquad (i \neq j; \quad i, j = 0, 1, \ldots, t).$$

To prove (3.33), we consider first the case $i = k$, where $s_{ij} \in \mathfrak{S}_{ij}$ implies $s_{ij}\mathfrak{A}_{ji} \subseteq \mathfrak{S}_{ii}$, the ideal of nilpotent elements in $\mathfrak{A}_{ii}$, implying $\mathfrak{S}_{ij}\mathfrak{A}_{ji} \subseteq \mathfrak{S}_{ii}$. Next, if $i \neq k$, we use (3.22) to see that

$$(\mathfrak{S}_{ij}\mathfrak{A}_{jk})\mathfrak{A}_{ki} = \mathfrak{S}_{ij}(\mathfrak{A}_{jk}\mathfrak{A}_{ki}) \subseteq \mathfrak{S}_{ij}\mathfrak{A}_{ji} \subseteq \mathfrak{S}_{ii}$$

by the result just established. But then $x_{ik}\mathfrak{A}_{ki}$ consists of nilpotent elements for all x_{ik} in $\mathfrak{S}_{ij}\mathfrak{A}_{jk} \subseteq \mathfrak{A}_{ik}$, establishing (3.33). To prove (3.34), we use (3.24) to see that

$$(\mathfrak{S}_{ij}\mathfrak{A}_{ij})\mathfrak{A}_{ij} = (\mathfrak{A}_{ij}\mathfrak{A}_{ij})\mathfrak{S}_{ij} \subseteq \mathfrak{A}_{ji}\mathfrak{S}_{ij} \subseteq \mathfrak{S}_{jj}$$

by the last statement before Lemma 3.6. Then $x_{ji}\mathfrak{A}_{ij}$ consists of nilpotent elements for all x_{ji} in $\mathfrak{S}_{ij}\mathfrak{A}_{ij} \subseteq \mathfrak{A}_{ji}$, implying (3.34). Then

$$\mathfrak{S}\mathfrak{A} = (\sum \mathfrak{S}_{ij})(\sum \mathfrak{A}_{kl}) = \sum \mathfrak{S}_{ij}\mathfrak{A}_{jk} + \sum_{i \neq j} \mathfrak{S}_{ij}\mathfrak{A}_{ij} \subseteq \mathfrak{S},$$

so that $\mathfrak{S}$ is a right ideal of $\mathfrak{A}$. Reciprocally, $\mathfrak{S}$ is a left ideal of $\mathfrak{A}$. We have proved that $\mathfrak{S} = \tilde{\mathfrak{P}}$ is an ideal of $\mathfrak{A}$.

Clearly the ideal $\tilde{\mathfrak{P}}$ is independent of the particular choice of idempotents $e_1, \ldots, e_t$ used above. Assume that $\tilde{\mathfrak{P}}$ is not a nilideal. Then $\tilde{\mathfrak{P}}$ contains an idempotent u ($\neq 0$) by Proposition 3.3. We have noted that there exist pairwise orthogonal primitive idempotents $e_1, \ldots, e_r, \ldots, e_t$ in $\mathfrak{A}$ such that $u = e_1 + \cdots + e_r$, while $e = e_1 + \cdots + e_t$ is principal. Then $u = e_1 + \cdots + e_r$ in $\tilde{\mathfrak{P}} = \mathfrak{S} = \sum \mathfrak{S}_{ij}$ implies e_1 in $\mathfrak{S}_{11}$ is nilpotent, a contradiction. Hence the nilideal $\tilde{\mathfrak{P}}$ is contained in $\mathfrak{N}$, and

$$\mathfrak{N} \subseteq \mathfrak{P} \subseteq \tilde{\mathfrak{P}} \subseteq \mathfrak{N},$$

implying $\mathfrak{N} = \mathfrak{P}$, as desired.

Corollary 3.8. *Let e be an idempotent in a finite-dimensional alternative algebra $\mathfrak{A}$ with radical $\mathfrak{N}$, and let (3.14) be the Peirce decomposition of $\mathfrak{A}$ relative to e. Then the radical of $\mathfrak{A}_{ii}$ is $\mathfrak{N} \cap \mathfrak{A}_{ii}$ ($i = 0, 1$).*

Proof. By Theorem 3.7 the radical $\mathfrak{N}_i$ of $\mathfrak{A}_{ii}$ consists of those elements of $\mathfrak{A}_{ii}$ which are properly nilpotent in $\mathfrak{A}_{ii}$. By (3.29) these elements are properly nilpotent in $\mathfrak{A}$, implying $\mathfrak{N}_i \subseteq \mathfrak{N} \cap \mathfrak{A}_{ii}$. Conversely, if $x_{ii} \in \mathfrak{A}_{ii}$ is in $\mathfrak{N}$, then x_{ii} is *a fortiori* in $\mathfrak{N}_i$. Hence $\mathfrak{N}_i = \mathfrak{N} \cap \mathfrak{A}_{ii}$.

Corollary 3.9. *Let e be a principal idempotent in a finite-dimensional alternative algebra $\mathfrak{A}$, and let (3.14) be the Peirce decomposition of $\mathfrak{A}$ relative to e. Then $\mathfrak{A}_{10} + \mathfrak{A}_{01} + \mathfrak{A}_{00}$ is contained in the radical $\mathfrak{N}$ of $\mathfrak{A}$.*

Proof. Using the notation $\mathfrak{S}_{ij}$ from the proof of Theorem 3.7, we have seen that

$$\mathfrak{A}_{10} = \sum_{i=1}^{t} \mathfrak{S}_{i0}, \qquad \mathfrak{A}_{01} = \sum_{j=1}^{t} \mathfrak{S}_{0j}, \qquad \mathfrak{A}_{00} = \mathfrak{S}_{00}.$$

Each of these subspaces is contained in $\tilde{\mathfrak{P}} = \mathfrak{N}$.

Theorem 3.10. *Every finite-dimensional semisimple alternative algebra $\mathfrak{A} \neq 0$ has an identity element 1.*

Proof. Since $\mathfrak{A}$ is not a nilalgebra, $\mathfrak{A}$ has a principal idempotent e. Let (3.14) be the Peirce decomposition of $\mathfrak{A}$ relative to e. By the Corollary immediately above, $\mathfrak{A}_{10} + \mathfrak{A}_{01} + \mathfrak{A}_{00} \subseteq \mathfrak{N} = 0$, so that $\mathfrak{A} = \mathfrak{A}_{11} = e\mathfrak{A}e$. That is, $e = 1$.

Corollary 3.11. *Every finite-dimensional simple alternative algebra $\mathfrak{A}$ has an identity element 1. The center $\mathfrak{C}$ of $\mathfrak{A}$ is a field, and $\mathfrak{A}$ is an (alternative central simple) algebra of finite dimension over $\mathfrak{C}$.*

Theorem 3.7 implies that every ideal $\mathfrak{B}$ in a finite-dimensional semi-simple alternative algebra $\mathfrak{A}$ is semisimple, while Theorem 3.10 gives a (1-1) correspondence between the idempotents in the center $\mathfrak{C}$ of $\mathfrak{A}$ and the nonzero ideals $\mathfrak{B}$ ($= \mathfrak{A}e$) of $\mathfrak{A}$. For suppose that z in $\mathfrak{B}$ is properly nilpotent in $\mathfrak{B}$. Then a in $\mathfrak{A}$ implies that aza is in $\mathfrak{B}$, so that $z(aza) = (za)^2$ is nilpotent. That is, za is nilpotent for all a in $\mathfrak{A}$, implying that z is properly nilpotent in $\mathfrak{A}$. Hence $z = 0$, and $\mathfrak{B}$ is semisimple by Theorem 3.7. If $\mathfrak{B} \neq 0$, let e be the identity element for $\mathfrak{B}$ given by Theorem 3.10. Then $\mathfrak{A}e \supseteq \mathfrak{B}e = \mathfrak{B}$ and $e\mathfrak{A} \supseteq e\mathfrak{B} = \mathfrak{B}$. On the other hand, $\mathfrak{A}e$ and $e\mathfrak{A}$ are contained in the ideal $\mathfrak{B}$. Hence $\mathfrak{B} = \mathfrak{A}e = e\mathfrak{A}$. For all a in $\mathfrak{A}$, we have

$$ae = e(ae) = (ea)e = ea,$$

and similarly (3.4) implies that, for all a, b in $\mathfrak{A}$, we have

$$(ae)b = (eae)b = e[a(eb)] = a(eb)$$

since $a(eb)$ is in $\mathfrak{B}$, so that $(a, e, b) = 0$. Hence e is in $\mathfrak{C}$. Conversely, for any idempotent e in $\mathfrak{C}$, the subspace $\mathfrak{B} = \mathfrak{A}e$ is a nonzero (semisimple) ideal of $\mathfrak{A}$:

$$\mathfrak{A}\mathfrak{B} = \mathfrak{A}(\mathfrak{A}e) = (\mathfrak{A}\mathfrak{A})e \subseteq \mathfrak{A}e = \mathfrak{B}$$

and

$$\mathfrak{B}\mathfrak{A} = (\mathfrak{A}e)\mathfrak{A} = \mathfrak{A}(e\mathfrak{A}) = \mathfrak{A}(\mathfrak{A}e) = \mathfrak{A}\mathfrak{B} \subseteq \mathfrak{B}.$$

It follows that, for any ideal $\mathfrak{B} \neq \mathfrak{A}, 0$, there is a complementary ideal $\mathfrak{B}_1$ such that $\mathfrak{A} = \mathfrak{B} \oplus \mathfrak{B}_1$. For, if $\mathfrak{B} = \mathfrak{A}e$, then $e_1 = 1 - e$ is an idempotent in $\mathfrak{C}$, and $\mathfrak{B}_1 = \mathfrak{A}e_1$ is the required ideal. Since both $\mathfrak{B}$ and $\mathfrak{B}_1$ are semisimple, the decomposition may be repeated until simple ideals are obtained. Finite-dimensionality insures that the process actually terminates.

Theorem 3.12 (Zorn). *A finite-dimensional alternative algebra $\mathfrak{A} \neq 0$ is semisimple if and only if $\mathfrak{A} = \mathfrak{S}_1 \oplus \cdots \oplus \mathfrak{S}_t$, for simple ideals $\mathfrak{S}_i$ $(i = 1, \ldots, t)$.*

Proof. We have shown the "only if" part above. Conversely, if $\mathfrak{A} = \mathfrak{S}_1 \oplus \cdots \oplus \mathfrak{S}_t$, then any nonzero ideal $\mathfrak{B}$ of $\mathfrak{A}$ is the direct sum of certain of the $\mathfrak{S}_i$, and $\mathfrak{S}_i^2 = \mathfrak{S}_i$ $(i = 1, \ldots, t)$ implies $\mathfrak{B}^2 = \mathfrak{B} \neq 0$. But then each algebra $\mathfrak{B}^{(j)}$ in the derived series for $\mathfrak{B}$ is $\mathfrak{B} \neq 0$, so $\mathfrak{B}$ cannot be solvable. That is, the radical $\mathfrak{N}$ of $\mathfrak{A}$ is 0, and the proof is complete.

If e_i is the identity element for $\mathfrak{S}_i$ $(i = 1, \ldots, t)$, then $1 = e_1 + \cdots + e_t$ is a sum of pairwise orthogonal idempotents e_i which are primitive in the center $\mathfrak{C}$ of $\mathfrak{A}$ (but not necessarily primitive in $\mathfrak{A}$). Also $\mathfrak{C} = \mathfrak{C}_1 \oplus \cdots \oplus \mathfrak{C}_t$ where $\mathfrak{C}_i$ is the center of $\mathfrak{S}_i$ $(i = 1, \ldots, t)$. By Theorem 3.12 and

Corollary 3.11, the structure of finite-dimensional semisimple alternative algebras is reduced to that of central simple algebras. In Theorem 3.17 we show that these are either associative or (8-dimensional) Cayley algebras.

If we were content to have Theorem 3.12 only for characteristic 0, we could have given a much shorter proof by applying Dieudonné's Theorem 2.6. For we have

Proposition 3.13. *The radical* $\mathfrak{N}$ *of any finite-dimensional alternative algebra* $\mathfrak{A}$ *over F of characteristic 0 is the radical* $\mathfrak{A}^{\perp}$ *of the trace form*

$$(x, y) = \text{trace } R_x R_y.$$

Proof. Clearly (x, y) is a symmetric bilinear form on $\mathfrak{A}$. Identities (3.2) imply that

$$R_{xy}R_z - R_x R_{yz} = [R_x, L_y]R_z + R_x[R_z, L_y] = [R_x R_z, L_y]$$

so that

$$(xy, z) - (x, yz) = \text{trace } R_{xy}R_z - \text{trace } R_x R_{yz} = 0$$

for all x, y, z in $\mathfrak{A}$. Hence $\mathfrak{A}^{\perp}$ is an ideal. If there were an idempotent e in $\mathfrak{A}^{\perp}$, then

$$0 = (e, e) = \text{trace } R_e{}^2 = \text{trace } R_e = \dim (\mathfrak{A}_{11} + \mathfrak{A}_{01}) \neq 0$$

for characteristic 0 by (3.14), a contradiction. Hence the nilideal $\mathfrak{A}^{\perp} \subseteq \mathfrak{N}$. If x is in $\mathfrak{N}$, then xy is nilpotent for all y in $\mathfrak{A}$, R_{xy} is nilpotent by (3.10), and

$$(x, y) = \text{trace } R_x R_y = \text{trace } R_{xy} + \text{trace}[L_y, R_x] = 0.$$

That is, x is in $\mathfrak{A}^{\perp}$, and we have $\mathfrak{N} \subseteq \mathfrak{A}^{\perp}$, $\mathfrak{N} = \mathfrak{A}^{\perp}$.

Hence $\mathfrak{A}$ (of characteristic 0) is semisimple if and only if (x, y) is nondegenerate. Therefore Theorem 2.6 yields the complicated "only if" part of Theorem 3.12 for F of characteristic 0.

4.   CAYLEY ALGEBRAS

We wish to define the class of Cayley algebras mentioned in the Introduction. We construct these algebras in the following manner. The procedure works slightly more smoothly if we assume that F has characteristic $\neq 2$. However, such an assumption would limit the entire remaining structure theory of alternative algebras to fields F of characteristic $\neq 2$. Since this restriction is unnecessary, we define Cayley algebras over an arbitrary field F, and prove the relevant facts at some cost in elegance. For some theorems, on which the structure theory does not depend, we shall later assume characteristic $\neq 2$.

An *involution* (*involutorial antiautomorphism*) of an algebra $\mathfrak{A}$ is a linear operator $x \to \bar{x}$ on $\mathfrak{A}$ satisfying

$$\overline{xy} = \bar{y}\bar{x}, \qquad \overline{\bar{x}} = x \qquad \text{for all} \quad x, y \text{ in } \mathfrak{A}.$$

Here we are concerned with an involution satisfying

$$x + \bar{x} \in F1, \qquad x\bar{x} \, (= \bar{x}x) \in F1 \qquad \text{for all} \quad x \text{ in } \mathfrak{A}.$$

Clearly this implies

(3.35) $$x^2 - t(x)x + n(x)1 = 0, \qquad t(x), \quad n(x) \text{ in } F,$$

with

(3.36) $\quad x + \bar{x} = t(x)1, \qquad x\bar{x} \, (= \bar{x}x) = n(x)1 \qquad \text{for all} \quad x \text{ in } \mathfrak{A}.$

Since $x \to \bar{x}$ is linear, the *trace* $t(x)$ is linear. We call $n(x)$ the *norm* of x. Since $\bar{1} = 1$, we have $t(\alpha 1) = 2\alpha$, $n(\alpha 1) = \alpha^2$ for all α in F.

Let $\mathfrak{B}$ be an algebra with 1 having dimension n over F such that $\mathfrak{B}$ has an involution $b \to \bar{b}$ satisfying (3.36). By the *Cayley–Dickson process* we construct an algebra $\mathfrak{A}$ of dimension $2n$ over F with the same properties and having $\mathfrak{B}$ as subalgebra (with 1 in $\mathfrak{B}$) as follows: $\mathfrak{A}$ consists of all ordered pairs $x = (b_1, b_2)$, b_i in $\mathfrak{B}$, addition and multiplication by scalars defined componentwise, and multiplication defined by

$$(b_1, b_2)(b_3, b_4) = (b_1 b_3 + \mu b_4 \bar{b}_2, \, \bar{b}_1 b_4 + b_3 b_2)$$

for all b_i in $\mathfrak{B}$ and some $\mu \neq 0$ in F. Then $1 = (1, 0)$ is a unity element for $\mathfrak{A}$, $\mathfrak{B}' = \{(b, 0) \mid b \in \mathfrak{B}\}$ is a subalgebra of $\mathfrak{A}$ isomorphic to $\mathfrak{B}$, $v = (0, 1)$ is an element of $\mathfrak{A}$ such that $v^2 = \mu 1$ and $\mathfrak{A}$ is the vector space direct sum $\mathfrak{A} = \mathfrak{B}' + v\mathfrak{B}'$ of the n-dimensional vector spaces $\mathfrak{B}'$, $v\mathfrak{B}'$. If we identify $\mathfrak{B}'$ with $\mathfrak{B}$, the elements of $\mathfrak{A}$ are of the form

(3.37) $\quad x = b_1 + vb_2 \qquad (b_i \text{ in } \mathfrak{B} \text{ uniquely determined by } x),$

and multiplication is given by

(3.38) $\quad (b_1 + vb_2)(b_3 + vb_4) = (b_1 b_3 + \mu b_4 \bar{b}_2) + v(\bar{b}_1 b_4 + b_3 b_2)$

for all b_i in $\mathfrak{B}$ and some $\mu \neq 0$ in F. Defining

$$\bar{x} = \bar{b}_1 - vb_2,$$

we have $\overline{xy} = \bar{y}\bar{x}$ since $b \rightarrow \bar{b}$ is an involution of $\mathfrak{B}$; hence $x \rightarrow \bar{x}$ is an involution of $\mathfrak{A}$. Also

$$x + \bar{x} = t(x)1, \qquad x\bar{x}(= \bar{x}x) = n(x)1$$

where, for x in (3.37), we have

(3.39) $$t(x) = t(b_1), \qquad n(x) = n(b_1) - \mu n(b_2).$$

When is $\mathfrak{A}$ alternative? Since $\mathfrak{A}$ is its own reciprocal algebra, it is sufficient to verify the left alternative law, which is equivalent to $(x, \bar{x}, y) = 0$ since

$$(x, \bar{x}, y) = (x, t(x)1 - x, y) = -(x, x, y).$$

Now

$$\begin{aligned}
(x, \bar{x}, y) &= n(x)y - (b_1 + vb_2)[(\bar{b}_1 b_3 - \mu b_4 \bar{b}_2) + v(b_1 b_4 - b_3 b_2)] \\
&= n(x)y - [b_1(\bar{b}_1 b_3) - \mu b_1(b_4 \bar{b}_2) + \mu(b_1 b_4)\bar{b}_2 - \mu(b_3 b_2)\bar{b}_2] \\
&\quad -v[\bar{b}_1(b_1 b_4) - \bar{b}_1(b_3 b_2) + (\bar{b}_1 b_3)b_2 - \mu(b_4 \bar{b}_2)b_2] \\
&= n(x)y - [n(b_1) - \mu n(b_2)](b_3 + vb_4) \\
&\quad - \mu(b_1, b_4, \bar{b}_2) - v(\bar{b}_1, b_3, b_2) \\
&= -\mu(b_1, b_4, \bar{b}_2) - v(\bar{b}_1, b_3, b_2)
\end{aligned}$$

by a trivial extension of Artin's Theorem 3.1. Hence $\mathfrak{A}$ is alternative if and only if $\mathfrak{B}$ is associative.

We shall have occasion to use the fact that, if $\mathfrak{A}$ is an alternative algebra with 1 and $\mathfrak{B}$ is a subalgebra with an involution $b \rightarrow \bar{b}$ satisfying (3.36), and if v in $\mathfrak{A}$ has the property

(3.40) $$bv = v\bar{b} \qquad \text{for all} \quad b \text{ in } \mathfrak{B}$$

and satisfies $v^2 = \mu 1$ (μ in F), then multiplication in $\mathfrak{B} + v\mathfrak{B}$ is given by (3.38). To substantiate this, we need only verify that

$$a(vb) = v(\bar{a}b),$$
$$(va)b = v(ba),$$
$$(va)(vb) = \mu b\bar{a}.$$

for all a, b in $\mathfrak{B}$. Now

$$0 = (v, \bar{a}, b) + (\bar{a}, v, b) = (v\bar{a})b - v(\bar{a}b) + (\bar{a}v)b - \bar{a}(vb)$$
$$= (v\bar{a})b - v(\bar{a}b) + (va)b - \bar{a}(vb) = [t(a)1 - \bar{a}]vb - v(\bar{a}b)$$
$$= a(vb) - v(\bar{a}b).$$

Similarly,

$$0 = (\bar{a}, \bar{b}, v) + (\bar{a}, v, \bar{b}) = (\bar{a}\bar{b})v - \bar{a}(\bar{b}v) + (\bar{a}v)\bar{b} - \bar{a}(v\bar{b})$$
$$= v(ba) + (\bar{a}v)[\bar{b} - t(b)1] = v(ba) - (va)b.$$

Finally,

$$(va)(vb) = (va)(\bar{b}v) = v(a\bar{b})v = v^2(b\bar{a}) = \mu b\bar{a}$$

by the Moufang identity (3.6).

Let the 2-dimensional algebra $\mathfrak{Z}$ over F be either $\mathfrak{Z} = F \oplus F$ or a separable quadratic field $\mathfrak{Z} = F(s)$ over F. In both cases, $\mathfrak{Z} = F1 + Fs$ where $s^2 = s + \alpha 1$, $4\alpha + 1 \neq 0$. In the latter case, $X^2 - X - \alpha$ is irreducible in $F[X]$. (If the characteristic of F is $\neq 2$, then $v_1 = s - \frac{1}{2}1$ satisfies $v_1{}^2 = \mu_1 1$, where $\mu_1 = (4\alpha + 1)/4 \neq 0$ in both cases, and μ_1 is a nonsquare in F if and only if $X^2 - X - \alpha$ is irreducible.) There is a unique involution, distinct from the identity operator, on $\mathfrak{Z}$. It satisfies (3.36). Hence we can use an iterative process (beginning with $\mathfrak{B} = \mathfrak{Z}$) to obtain by the above construction algebras of dimension 2^t over F; these depend completely upon α and the $t - 1$ nonzero scalars $\mu_2, \ldots, \mu_t$ used in the successive steps. (If the characteristic of F is $\neq 2$, we obtain identical results by beginning with $F1$ on which the identity operator is an involution satisfying (3.36). Here the 2^t-dimensional algebras depend completely on the nonzero scalars $\mu_1, \mu_2, \ldots, \mu_t$.)

The 4-dimensional algebras $\mathfrak{Q} = \mathfrak{Z} + v_2 \mathfrak{Z}$ obtained in this fashion are associative central simple algebras (called *quaternion algebras*) over F. Any $\mathfrak{Q}$ which is not a division algebra is (by Wedderburn's theorem on simple associative algebras) isomorphic to the algebra of all 2×2 matrices with elements in F.

We are concerned with the 8-dimensional algebras $\mathfrak{C} = \mathfrak{Q} + v_3 \mathfrak{Q}$ which are called *Cayley algebras* over F. Since any $\mathfrak{Q}$ is associative, Cayley algebras are alternative. However, no Cayley algebra is associative. For $\mathfrak{Q}$ is not commutative and there exist q_1, q_2 in $\mathfrak{Q}$ such that $[q_1, q_2] \neq 0$; hence

$$(v_3, q_2, q_1) = (v_3 q_2)q_1 - v_3(q_2 q_1) = v_3[q_1, q_2] \neq 0.$$

Thus this iterative process of constructing alternative algebras stops after three steps. A multiplication table for any $\mathfrak{C}$ of characteristic $\neq 2$ is given on page 5.

A Cayley algebra $\mathfrak{C}$ is a division algebra if and only if $n(x) \neq 0$ for every $x \neq 0$ in $\mathfrak{C}$. For $x \neq 0$, $n(x) = 0$ imply $x\bar{x} = n(x)1 = 0$, $\mathfrak{C}$ has zero divisors. Conversely, if $n(x) \neq 0$, then $\bar{x}(xy) = (\bar{x}x)y = n(x)y$ for all y implies

$$\frac{1}{n(x)} L_x L_{\bar{x}} = 1_{\mathfrak{C}}, \qquad L_x^{-1} = \frac{1}{n(x)} L_{\bar{x}}$$

and, similarly,

$$R_x^{-1} = \frac{1}{n(x)} R_{\bar{x}};$$

hence, if $n(x) \neq 0$ for all $x \neq 0$, $\mathfrak{C}$ is a division algebra.

Remark. If F is the field of all real numbers, the norm form $n(x) = \sum \alpha_i^2$ for $x = \sum \alpha_i u_i$ clearly has the property above. Also there are alternative algebras $F1$, $\mathfrak{Z}$, $\mathfrak{Q}$, $\mathfrak{C}$ with this norm form (take $\mu_i = -1$ at each step). Hence there are real alternative division algebras of dimensions 1, 2, 4, 8. In 1958 it was proved that finite-dimensional real division algebras can have only these dimensions (Bott and Milnor [1]). It is not true, however, that the only finite-dimensional real division algebras are the four listed above; they are the only alternative ones. See Bruck [1] and Osborn [1] for other examples of finite-dimensional real division algebras (necessarily of these specified dimensions, of course).

Over any field F there is a Cayley algebra with divisors of zero (take $\mu = 1$ so $v^2 = 1$). This Cayley algebra over F, which we shall prove in Lemma 3.16 is unique, is called the *split Cayley algebra* over F.

$F1$ is both the nucleus and center of any Cayley algebra. Also any Cayley algebra is simple (hence central simple over F); this is obvious for Cayley division algebras, but not for the split Cayley algebra. If $\mathfrak{B}$ is any nonzero ideal of $\mathfrak{C} = \mathfrak{Q} + v\mathfrak{Q}$, there is $x = a + vb \neq 0$ in $\mathfrak{B}$ (a, b in $\mathfrak{Q}$). If $a = 0$, then $\mathfrak{Q}(vx)\mathfrak{Q}$ is a nonzero ideal of the simple algebra $\mathfrak{Q}$,

$$1 \in \mathfrak{Q} = \mathfrak{Q}(vx)\mathfrak{Q} \subseteq \mathfrak{B},$$

and $\mathfrak{B} = \mathfrak{C}$. If $a \neq 0$, then

$$1 \in \mathfrak{Q}a\mathfrak{Q} \subseteq \mathfrak{Q}x\mathfrak{Q} + \mathfrak{Q}(vb)\mathfrak{Q} \subseteq \mathfrak{B} + v\mathfrak{Q},$$

implying $1 - vh$ is in $\mathfrak{B}$ for some h in $\mathfrak{Q}$. Now $\mathfrak{Z} = F1 + Fs$ with $s^2 = s + \alpha 1$, so $s - \beta\bar{s} = -\beta 1 + (1 + \beta) s \neq 0$ for all β in F. Then $\mathfrak{B}$ contains

$$(1 - vh)s + [v(sh)](1 - vh) = s - \mu n(h)\bar{s} = q \neq 0$$

in $\mathfrak{Q}$, and $1 \in \mathfrak{Q}q\mathfrak{Q} \subseteq \mathfrak{B}$, so $\mathfrak{B} = \mathfrak{C}$ in this case also.

Consider an algebra $\mathfrak{A}$ with 1 (of possibly infinite dimension) over an arbitrary field F. Assume that for each x in $\mathfrak{A}$ we have

$$x^2 - t(x)x + n(x)1 = 0, \qquad t(x), n(x) \text{ in } F,$$

that is, (3.35) holds in $\mathfrak{A}$, but no further properties are assumed. If x is not in $F1$, the scalars $t(x)$, $n(x)$ are uniquely determined. Put

$$t(\alpha 1) = 2\alpha, \qquad n(\alpha 1) = \alpha^2 \qquad \text{for all } \alpha \text{ in } F,$$

so that the *trace* $t(x)$ and the *norm* $n(x)$ are unique in (3.35) for all x in $\mathfrak{A}$. Now (3.35) implies $t(\alpha x) = \alpha t(x)$ for all α in F, x in $\mathfrak{A}$. If F contains more than two elements, then the trace $t(x)$ is linear (whereas an 8-element Boolean algebra with 1 satisfies (3.35), but has elements a, b such that $t(a + b) = 1 \neq t(a) + t(b)$). We wish to show that the trace is additive:

(3.41) $\qquad t(x + y) = t(x) + t(y) \qquad$ for all $\quad x, y$ in $\mathfrak{A}$.

It is easy to see that (3.35) implies (3.41) if $y = \beta 1$, β in F. It follows that (3.41) is satisfied if $x, y, 1$ are linearly dependent. Therefore we can assume $x, y, 1$ are linearly independent, and compute

$$0 = (\alpha x + \beta y)^2 - t(\alpha x + \beta y)(\alpha x + \beta y) + n(\alpha x + \beta y)1$$
$$- \alpha\beta[(x + y)^2 - t(x + y)(x + y) + n(x + y)1]$$

for all α, β in F, as a linear combination of $x, y, 1$ to obtain

$$\alpha[\alpha t(x) - \beta t(x) - t(\alpha x + \beta y) + \beta t(x + y)] = 0,$$

and

$$\beta[\beta t(y) - \alpha t(y) - t(\alpha x + \beta y) + \alpha t(x + y)] = 0.$$

If $\alpha \neq 0$ and $\beta \neq 0$, we may divide by these scalars, and subtract, to obtain

$$(\alpha - \beta)[t(x) + t(y) - t(x + y)] = 0.$$

Hence, if $0, \alpha, \beta$ are distinct in F, we have (3.41). In order to facilitate the passage to scalar extensions, we wish to have a linear trace. We are led therefore to the following definition.

Let $\mathfrak{A} \neq F1$ be an algebra with 1 over F such that for each x in $\mathfrak{A}$ we have

$$x^2 - t(x)x + n(x)1 = 0, \qquad t(x), \, n(x) \text{ in } F;$$

in addition, if F is the field of two elements, the trace $t(x)$ (uniquely defined by setting $t(\alpha 1) = 2\alpha$) is required to be linear. Then $\mathfrak{A}$ is called a *quadratic algebra* over F. If $\mathfrak{A}$ is a quadratic algebra, then so is any scalar extension $\mathfrak{A}_K$. This may easily be seen by using a basis for $\mathfrak{A}$ and the linearity of the trace on $\mathfrak{A}$.

Conversely, if $\mathfrak{A}$ is an algebra with 1 over F such that $\mathfrak{A}_K$ is a quadratic algebra over K, then $\mathfrak{A}$ is a quadratic algebra. For x in $\mathfrak{A}$, we have (3.35) with $t(x), \, n(x)$ in K. We need to show that $t(x)$ and $n(x)$ are in F. Writing $u_0 = 1$, we let $\{u_i\}$ be a basis for $\mathfrak{A}$ over F. Then

$$u_i u_j = \sum \gamma_{ijk} u_k, \qquad \gamma_{ijk} \text{ in } F.$$

Also $\{u_i\}$ is a basis for $\mathfrak{A}_K$ over K. For $h \neq 0$, we have

$$u_h^2 = t(u_h)u_h - n(u_h)u_0 = \sum \gamma_{hhk} u_k,$$

implying $t(u_h) = \gamma_{hhh}$ is in F $(h \neq 0)$. Then $x = \alpha_0 u_0 + \sum_{h \neq 0} \alpha_h u_h$ implies

$$t(x) = 2\alpha_0 + \sum \alpha_h t(u_h)$$

is in F, since the trace is linear on $\mathfrak{A}_K$. Then $t(x)$ is linear on $\mathfrak{A}$. Also

$$x^2 = \sum \alpha_i \alpha_j \gamma_{ijk} u_k = t(x) \sum \alpha_i u_i - n(x)u_0$$

implies that

$$n(x) = t(x)\alpha_0 - \sum \alpha_i \alpha_j \gamma_{ij0}$$

is in F.

All of the algebras $(\neq F1)$ which are constructed by the Cayley–Dickson process are quadratic algebras over an arbitrary field F.

5.   SIMPLE ALTERNATIVE ALGEBRAS

We assume now that $\mathfrak{A}$ is a finite-dimensional simple alternative algebra over an arbitrary field F. Then $\mathfrak{A}$ contains 1. If 1 is a primitive idempotent in $\mathfrak{A}$, then Lemma 3.5 implies that $\mathfrak{A}$ is a division algebra. More generally, we may consider, as in the proofs of Lemma 3.6 and

Theorem 3.7, the Peirce decomposition of $\mathfrak{A}$ relative to pairwise orthogonal idempotents $e_1, \ldots, e_t$ such that $1 = e_1 + \cdots + e_t$. It follows from Corollary 3.8 that $\mathfrak{A}_{ii}$ is semisimple $(i = 1, \ldots, t)$. Applying Lemma 3.5 to $\mathfrak{A}_{ii}$, we see that $\mathfrak{A}_{ii}$ is a division algebra $(i = 1, \ldots, t)$. If $t > 1$, it follows from Lemma 3.14 that $\mathfrak{A}_{ii}$ is associative (a "skew field").

Lemma 3.14. *Let $\mathfrak{A}$ be a finite-dimensional simple alternative algebra with idempotent $e \neq 1$. Let (3.14) be the Peirce decomposition of $\mathfrak{A}$ relative to e. Then $e\mathfrak{A}e = \mathfrak{A}_{10}\mathfrak{A}_{01}$ is associative.*

Proof. $\mathfrak{A}_{10}\mathfrak{A}_{01}$ (resp. $\mathfrak{A}_{01}\mathfrak{A}_{10}$) is an ideal of $\mathfrak{A}_{11}$ (resp. $\mathfrak{A}_{00}$) by (3.17) and (3.22). Hence

$$\mathfrak{B} = \mathfrak{A}_{10}\mathfrak{A}_{01} + \mathfrak{A}_{10} + \mathfrak{A}_{01} + \mathfrak{A}_{01}\mathfrak{A}_{10}$$

is an ideal of $\mathfrak{A}$ by (3.17)–(3.19). If $\mathfrak{B} = 0$, then $\mathfrak{A} = \mathfrak{A}_{11} \oplus \mathfrak{A}_{00}$ with $\mathfrak{A}_{11} \neq 0$, $\mathfrak{A}_{00} \neq 0$, a contradiction. Hence $\mathfrak{B} = \mathfrak{A}$, and $e\mathfrak{A}e = \mathfrak{A}_{11} = \mathfrak{A}_{10}\mathfrak{A}_{01}$. Also (3.23) implies that $\mathfrak{A}_{10}\mathfrak{A}_{01}$ is associative.

Lemma 3.15. *Let $\mathfrak{A}$ be a finite-dimensional simple alternative algebra, and let $1 = e_1 + \cdots + e_t$ for pairwise orthogonal idempotents e_i $(i = 1, \ldots, t)$. If $t \geq 3$, then $\mathfrak{A}$ is associative.*

Proof. Let (3.15) be the Peirce decomposition relative to $e_1, \ldots, e_t$. For any $i = 2, \ldots, t$, we wish to show first that

$$\mathfrak{A}_{1i}^2 = \mathfrak{A}_{i1}^2 = 0 \qquad (i = 2, \ldots, t).$$

Let $e = e_1 + e_i \neq 1$ since $t \geq 3$. Lemma 3.14 implies that $e\mathfrak{A}e = \mathfrak{A}_{11} + \mathfrak{A}_{1i} + \mathfrak{A}_{i1} + \mathfrak{A}_{ii}$ is associative. But this is the Peirce decomposition of $e\mathfrak{A}e$ relative to e_1, and we have $\mathfrak{A}_{1i}^2 = \mathfrak{A}_{i1}^2 = 0$. It follows from (3.19) that

$$\left(\sum_{j=2}^{t} \mathfrak{A}_{1j} \right)^2 = \left(\sum_{i=2}^{t} \mathfrak{A}_{i1} \right)^2 = 0.$$

Consider now the Peirce decomposition (3.14) relative to the idempotent e_1. (There is no ambiguity about the notation $\mathfrak{A}_{11}$.) It follows from (3.17), (3.19) and the fact that associators alternate that $\mathfrak{A}$ is associative in case $\mathfrak{A}_{11}$ and $\mathfrak{A}_{00}$ are associative and $\mathfrak{A}_{10}^2 = \mathfrak{A}_{01}^2 = 0$. But

$$\mathfrak{A}_{10} = \sum_{j=2}^{t} \mathfrak{A}_{1j}, \qquad \mathfrak{A}_{01} = \sum_{i=2}^{t} \mathfrak{A}_{i1}, \qquad \mathfrak{A}_{00} = \sum_{i,j=2}^{t} \mathfrak{A}_{ij}.$$

Hence $\mathfrak{A}_{10}^2 = \mathfrak{A}_{01}^2 = 0$. We already know that $\mathfrak{A}_{11}$ is associative and, applying Lemma 3.14 to the case $e = e_2 + \cdots + e_t$, we have $\mathfrak{A}_{00}$ associative. Hence $\mathfrak{A}$ is associative.

Lemma 3.16. *Let $\mathfrak{A}$ be a finite-dimensional simple alternative algebra over (an arbitrary field) F satisfying:*

(i) $1 = e_1 + e_2$ *for (orthogonal) primitive idempotents e_i;*
(ii) $\mathfrak{A}_{ii} (= e_i \mathfrak{A} e_i) = Fe_i$ $(i = 1, 2)$;
(iii) $\mathfrak{A}$ *is not associative.*

Then $\mathfrak{A}$ is the unique split Cayley algebra over F: $\mathfrak{A} = \mathfrak{Q} + v\mathfrak{Q}$ where $\mathfrak{Q} \cong F_2$, the algebra of all 2×2 matrices over F, and multiplication in $\mathfrak{A}$ is defined by (3.38) with $\mu = 1$.

Proof.   Let $\mathfrak{B}$ be any semisimple subalgebra of $\mathfrak{A}$ which contains e_1 and e_2. (Note that $\mathfrak{A}$ is such a subalgebra $\mathfrak{B}$.) Let

$$\mathfrak{B} = \mathfrak{B}_{11} + \mathfrak{B}_{12} + \mathfrak{B}_{21} + \mathfrak{B}_{22}$$

be the Peirce decomposition (3.15) of $\mathfrak{B}$ relative to e_1, e_2. Then $\mathfrak{B}_{ij} \subseteq \mathfrak{A}_{ij}$. Also (ii) implies

$$\mathfrak{B} = Fe_1 + \mathfrak{B}_{12} + \mathfrak{B}_{21} + Fe_2.$$

Now x_{12} in $\mathfrak{B}_{12}$, y_{21} in $\mathfrak{B}_{21}$ imply $x_{12} y_{21}$ is in $\mathfrak{B}_{11} = Fe_1$, so that

$$x_{12} y_{21} = \langle x_{12}, y_{21} \rangle e_1 \qquad \text{for all} \quad x_{12} \text{ in } \mathfrak{B}_{12}, \quad y_{21} \text{ in } \mathfrak{B}_{21},$$

where $\langle x, y \rangle$ is a bilinear form connecting $\mathfrak{B}_{12}$ and $\mathfrak{B}_{21}$. (By the parenthetical remark above there is a bilinear form connecting $\mathfrak{A}_{12}$ and $\mathfrak{A}_{21}$ which induces $\langle x, y \rangle$.) It follows from the flexible law that

$$y_{21} x_{12} = \langle x_{12}, y_{21} \rangle e_2 \qquad \text{for all} \quad x_{12} \text{ in } \mathfrak{B}_{12}, \quad y_{21} \text{ in } \mathfrak{B}_{21}.$$

For this certainly holds if $y_{21} = 0$. Taking $y_{21} \neq 0$, we have $y_{21} x_{12} = \alpha e_2$, $\alpha \in F$, since $y_{21} x_{12}$ is in $\mathfrak{B}_{22} = Fe_2$. Then

$$\alpha y_{21} = \alpha e_2 y_{21} = y_{21} x_{12} y_{21} = \langle x_{12}, y_{21} \rangle y_{21},$$

implying $\alpha = \langle x_{12}, y_{21} \rangle$. Now the left radical of $\langle x, y \rangle$ is 0: $\langle x_{12}, y_{21} \rangle = 0$ for all y_{21} in $\mathfrak{B}_{21}$ implies $x_{12} = 0$. For, if $x_{12} \neq 0$, then (3.32) implies the existence of b_{21} in $\mathfrak{B}_{21}$ satisfying

(3.42)
$$x_{12} b_{21} = e_1, \qquad b_{21} x_{12} = e_2,$$

and we have $e_1 = x_{12} b_{21} = \langle x_{12}, b_{21} \rangle e_1 = 0$, a contradiction. Similarly, for any $x_{21} \neq 0$ in $\mathfrak{B}_{21}$, we have b_{12} in $\mathfrak{B}_{12}$ satisfying

$$b_{12} x_{21} = e_1, \qquad x_{21} b_{12} = e_2,$$

implying that the right radical of $\langle x, y \rangle$ is 0. Hence $\langle x, y \rangle$ is nondegenerate and dim $\mathfrak{B}_{12} = $ dim $\mathfrak{B}_{21}$ (Jacobson [24], p. 141; Artin [2], p. 21). In particular, dim $\mathfrak{A}_{12} = $ dim $\mathfrak{A}_{21}$.

If $\mathfrak{B} \neq \mathfrak{A}$, then $k = $ dim $\mathfrak{B}_{12} < $ dim $\mathfrak{A}_{12}$, since

$$2 + 2k = \dim \mathfrak{B} < \dim \mathfrak{A} = 2 + 2 \dim \mathfrak{A}_{12}.$$

Hence there exists $f_{12} \in \mathfrak{A}_{12}, f_{12} \notin \mathfrak{B}_{12}$. Let $u_1, \ldots, u_k$ be a basis for $\mathfrak{B}_{12}$. Since $\langle x, y \rangle$ is nondegenerate, there is a dual basis $w_1, \ldots, w_k$ for $\mathfrak{B}_{21}$ satisfying

(3.43)
$$\langle u_i, w_j \rangle = \delta_{ij} \qquad (i, j = 1, \ldots, k)$$

(Jacobson [24], p. 142; Artin [2], p. 18). Now the element

$$g_{12} = f_{12} - \sum_{i=1}^{k} \langle f_{12}, w_i \rangle u_i$$

is in $\mathfrak{A}_{12}$, but g_{12} is not in $\mathfrak{B}_{12}$. Then

(3.44)
$$\langle g_{12}, w_j \rangle = 0 \qquad (j = 1, \ldots, k).$$

Since (3.42) holds (in particular) for $\mathfrak{A}$, we have g_{21} in $\mathfrak{A}_{21}$ satisfying

$$g_{12} g_{21} = e_1, \qquad g_{21} g_{12} = e_2.$$

Let

$$h_{21} = g_{21} - \sum_{j=1}^{k} \langle u_j, g_{21} \rangle w_j \qquad \text{in } \mathfrak{A}_{21}.$$

Then

$$g_{12} h_{21} = e_1, \qquad h_{21} g_{12} = e_2$$

by (3.44). Also

(3.45)
$$\langle u_i, h_{21} \rangle = 0 \qquad (i = 1, \ldots, k)$$

by (3.43). Let $v = g_{12} + h_{21}$. Then v is not in $\mathfrak{B}$ since g_{12} is not in $\mathfrak{B}_{12}$. Also

$$v^2 = 1.$$

For $v^2 = (g_{12} + h_{21})^2 = e_1 + e_2 = 1$ by (3.20).

Assume that $\mathfrak{B}$ has an involution $b \to \bar{b}$ satisfying (3.36). [For example, $Fe_1 \oplus Fe_2$ is a semisimple subalgebra $\mathfrak{B}$ with involution

$$b = \alpha e_1 + \beta e_2 \to \bar{b} = \beta e_1 + \alpha e_2$$

satisfying (3.36).] Any b in $\mathfrak{B}$ may be written in the form

$$b = \alpha e_1 + \sum \alpha_i u_i + \sum \beta_j w_j + \beta e_2; \qquad \alpha, \beta, \alpha_i, \beta_j \text{ in } F.$$

It follows from (3.35) and (3.20) that

$$\bar{u}_i = -u_i, \qquad \bar{w}_i = -w_i \qquad (i = 1, \dots, k).$$

Also (3.35) implies

$$\bar{e}_1 = e_2, \qquad \bar{e}_2 = e_1.$$

Hence

$$\bar{b} = \beta e_1 - \sum \alpha_i u_i - \sum \beta_j w_j + \alpha e_2.$$

It follows from (3.21), (3.44), and (3.45) that

$$\begin{aligned} bv &= \alpha g_{12} + \sum \alpha_i u_i \, g_{12} + \sum \beta_j w_j h_{21} + \beta h_{21} \\ &= \beta h_{21} - \sum \alpha_i g_{12} u_i - \sum \beta_j h_{21} w_j + \alpha g_{12} \\ &= v\bar{b}. \end{aligned}$$

That is, (3.40) holds for all b in $\mathfrak{B}$. Hence we have (3.38) with $\mu = 1$, and $\mathfrak{B} + v\mathfrak{B}$ is a homomorphic image in $\mathfrak{A}$ of the $(2n)$-dimensional algebra constructed by the Cayley–Dickson process from the algebra $\mathfrak{B}$ of dimension $n = 2 + 2k$. Beginning with the 2-dimensional algebra $Fe_1 \oplus Fe_2$, the process does not terminate until we have $\mathfrak{B} = \mathfrak{A}$. By (iii) $\mathfrak{A} \neq Fe_1 \oplus Fe_2$. The case $k = 0$ gives a homomorphic image in $\mathfrak{A}$ of the (simple) algebra F_2 of all 2×2 matrices over F. Since the homomorphic image is $\neq 0$, the homomorphism is a monomorphism, and $\mathfrak{A}$ contains

$\mathfrak{Q} \cong F_2$. By (iii) $\mathfrak{A} \neq \mathfrak{Q}$. The case $k = 1$ gives a homomorphic image in $\mathfrak{A}$ of the (simple) Cayley algebra described in the conclusion of this lemma; hence $\mathfrak{A}$ contains a Cayley subalgebra $\mathfrak{C}$. If $\mathfrak{A} \neq \mathfrak{C}$, we have shown (in the case $k = 3$) the existence of $v \notin \mathfrak{C}$ satisfying (3.38) with $\mu = 1$ for all b_i in $\mathfrak{C}$. Choose b_i ($i = 1, 2, 3$) in $\mathfrak{C}$ such that $(b_1, b_2, b_3) \neq 0$. We have shown earlier that (3.38) implies

$$(b_1 + vb_2, b_1 + vb_2, b_3) = -v(b_1, b_3, b_2).$$

Hence

$$0 = v[v(b_1, b_3, b_2)] = -(b_1, b_2, b_3) \neq 0,$$

a contradiction. That is, $\mathfrak{A} = \mathfrak{C}$, as desired.

We have shown that any $\mathfrak{A}$ satisfying (i)–(iii) is the unique Cayley algebra described in the conclusion. Since $e_1 e_2 = 0$, this is a Cayley algebra with divisors of zero. To prove that there is only one Cayley algebra with divisors of zero over an arbitrary field F, we need only observe that conditions (i)–(iii) are satisfied in any such algebra $\mathfrak{A}$. Clearly (iii) is true. Write $1 = e_1 + \cdots + e_t$ for pairwise orthogonal primitive idempotents e_i in $\mathfrak{A}$. Then (i) holds, since we have seen (preceding Lemma 3.14) that $\mathfrak{A}$ is a division algebra if $t = 1$, and is associative by Lemma 3.15 if $t \geq 3$. Let K be the algebraic closure of F. Then 1 is the identity element for $\mathfrak{A}_K$. Since the Cayley algebra $\mathfrak{A}$ is central simple, $\mathfrak{A}_K$ is simple. In particular, the radical of $\mathfrak{A}_K$ is 0. It follows from Corollary 3.8 that therefore the radical of $(\mathfrak{A}_K)_{ii} = (\mathfrak{A}_{ii})_K$ is 0 ($i = 1, 2$). We have seen (as a consequence of Lemma 3.14) that, since $t > 1$, each $\mathfrak{A}_{ii}$ is associative. Hence $(\mathfrak{A}_{ii})_K$ is a direct sum of simple associative algebras over K. Since K is algebraically closed, these are (by Wedderburn's theorem) total matrix algebras over K. If there is more than one summand in $(\mathfrak{A}_{ii})_K$, or if that summand is not Ke_i, then $1 = e_1 + e_2$ is expressible as a sum of more than two pairwise orthogonal idempotents in $\mathfrak{A}_K$, and $\mathfrak{A}_K$ is associative by Lemma 3.15, $\mathfrak{A}$ ($\subseteq \mathfrak{A}_K$) is associative, a contradiction. Hence $(\mathfrak{A}_{ii})_K = Ke_i$ ($i = 1, 2$), implying (ii). That is, over an arbitrary field F there is one and (up to isomorphism) only one split Cayley algebra.

One can easily construct over an arbitrary field F an explicit isomorphism between Zorn's "vector matrix algebra" (Jacobson [25], p.

142) and the split Cayley algebra given in Lemma 3.16.

Theorem 3.17 (Zorn). *The finite-dimensional central simple alternative algebras over F are the 8-dimensional Cayley algebras over F and the $(mn)^2$-dimensional algebras $\mathfrak{D}_n = \mathfrak{D} \otimes F_n$, $\mathfrak{D}$ a central associative division algebra of degree m over F.*

Proof. By Wedderburn's theorem on finite-dimensional simple associative algebras it remains only to prove that any finite-dimensional central simple alternative algebra $\mathfrak{A}$ which is not associative is a Cayley algebra. Writing $1 = e_1 + \cdots + e_t$ for pairwise orthogonal primitive idempotents e_i in $\mathfrak{A}$, we have $t \leq 2$ by Lemma 3.15. If $t = 1$, we know that $\mathfrak{A}$ is a division algebra. Let K be the algebraic closure of F. The decomposition of 1 above as a sum of pairwise orthogonal idempotents in $\mathfrak{A}$ may be refined to one in $\mathfrak{A}_K$: $1 = e_1' + \cdots + e_{t'}'$. Then $t' \leq 2$ since $\mathfrak{A}_K$ is simple but not associative. If $t' = 1$, then $\mathfrak{A}_K = (\mathfrak{A}_K)_{11}$ is a finite-dimensional alternative division algebra over K. But then any x in $\mathfrak{A}_K$ generates a subfield $K(x)$ of $\mathfrak{A}_K$, where $K(x)$ is of finite degree over the algebraically closed field K. Hence $K(x) = K$, x is in K. That is, $\mathfrak{A}_K = K1$, $\mathfrak{A} = F1$, a contradiction. Hence $t' = 2$. That is, there are two possible situations: (a) $\mathfrak{A}$ is a division algebra, but $1 = e_1 + e_2$ for primitive idempotents e_i in $\mathfrak{A}_K$, and (b) $1 = e_1 + e_2$ for primitive idempotents e_i in $\mathfrak{A}$ which remain primitive in $\mathfrak{A}_K$ (such e_i are called *absolutely primitive*). In case (b) we have the situation which was considered in the final paragraph of the proof of Lemma 3.16: $\mathfrak{A}$ is the split Cayley algebra over F. In case (a) the algebra $\mathfrak{A}_K$ satisfies the conditions for case (b): $\mathfrak{A}_K$ is the split Cayley algebra over K. In particular, $\mathfrak{A}_K$ is a quadratic algebra over K. It remains to show that this implies in (a) that the division algebra $\mathfrak{A}$ is a Cayley algebra.

Since $\mathfrak{A}_K$ is a quadratic algebra over K, we have seen that $\mathfrak{A}$ is a quadratic algebra over F. Now $\mathfrak{A}$ is not commutative, since the Cayley algebra $\mathfrak{A}_K$ is not. Take any pair x, y of elements in $\mathfrak{A}$ which do not commute. By Artin's Theorem 3.1 they generate an associative, but not commutative, subalgebra $\mathfrak{Q}$ of $\mathfrak{A}$. Since $\mathfrak{Q}$ contains no zero divisors, finite-dimensionality ensures that $\mathfrak{Q}$ is a division subalgebra of $\mathfrak{A}$. Also 1 is in $\mathfrak{Q}$. Since $\mathfrak{Q} \subseteq \mathfrak{A}$ is a quadratic algebra over F, it is known that $\mathfrak{Q}$ is a quaternion division algebra (Albert [24], p. 146). It remains only to show the existence of $v \in \mathfrak{A}$, $v \notin \mathfrak{Q}$, satisfying

$$(3.46) \qquad v^2 = \mu 1, \qquad \mu \neq 0 \text{ in } F,$$

and (3.40) for all b in $\mathfrak{Q}$.

Now $\mathfrak{Q} = \mathfrak{Z} + v_2 \mathfrak{Z}$, $\mathfrak{Z} = F(s)$, $s^2 = s + \alpha 1$, $4\alpha + 1 \neq 0$. Hence $\mathfrak{Q}$ has a basis $1, s, v_2, v_2 s$ over F and multiplication table:

	1	s	v_2	$v_2 s$
1	1	s	v_2	$v_2 s$
s	s	$\alpha 1 + s$	$v_2 - v_2 s$	$-\alpha v_2$
v_2	v_2	$v_2 s$	$\mu_2 1$	$\mu_2 s$
$v_2 s$	$v_2 s$	$\alpha v_2 + v_2 s$	$\mu_2 1 - \mu_2 s$	$-\mu_2 \alpha 1$

Since $\mathfrak{Q} \neq \mathfrak{A}$, there exists $f \in \mathfrak{A}$, $f \notin \mathfrak{Q}$. Let

$$v = f + \lambda_1 1 + \lambda_2 s + \lambda_3 v_2 + \lambda_4 v_2 s$$

with λ_i in F. Then v is not in $\mathfrak{Q}$. We seek λ_i in F for which (3.46) and (3.40) are satisfied. By (3.35) we see that (3.46) is equivalent to $t(v) = 0$, or

$$(3.47) \qquad 2\lambda_1 + \lambda_2 = -t(f),$$

since $\mu \neq 0$ in the division algebra $\mathfrak{A}$. Now the involution in $\mathfrak{A}_K$ induces $x \to \bar{x} = t(x)1 - x$ in $\mathfrak{A}$, which is therefore an involution in $\mathfrak{A}$. Assuming $t(v) = 0$, we have $\bar{v} = -v$ so that

$$t(bv)1 = bv + \overline{bv} = bv + \bar{v}\bar{b} = bv - v\bar{b},$$

implying that (3.40) is equivalent to $t(bv) = 0$ for all b in $\mathfrak{Q}$; that is, to $t(v) = t(sv) = t(v_2 v) = t((v_2 s)v) = 0$. Hence we seek λ_i in F satisfying (3.47) and

$$\lambda_1 + (1 + 2\alpha)\lambda_2 = -t(sf),$$

$$2\mu_2 \lambda_3 + \mu_2 \lambda_4 = -t(v_2 f),$$

$$\mu_2 \lambda_3 - 2\alpha\mu_2 \lambda_4 = -t((v_2 s)f).$$

Since the determinant of the four equations is $-\mu_2^2(1 + 4\alpha)^2 \neq 0$, the desired λ_i in F exist. Hence $\mathfrak{A}$ is a Cayley division algebra in case (a) above.

The proof that any simple alternative ring, which is not a nilring and which is not associative, is a Cayley algebra over its center appears in

Kleinfeld [2]. Of geometric significance is the case of division rings, the easier proof for which (assuming characteristic $\neq 2$) is given in Kleinfeld [9].

The norm $n(x)$ on any involutorial alternative algebra $\mathfrak{A}$ satisfying (3.36) *permits composition* in the sense that

(3.48) $n(xy) = n(x)n(y)$ for all x, y in $\mathfrak{A}$.

For

$$n(xy)1 = (xy)(\overline{xy}) = xy\bar{y}\bar{x} = n(y)x\bar{x} = n(x)n(y)1$$

by Artin's Theorem 3.1.

We remark that a necessary and sufficient condition that a Cayley algebra $\mathfrak{C} = \mathfrak{Q} + v\mathfrak{Q}$ be a division algebra is that we have

$$n(x) = n(b_1) - \mu n(b_2) \neq 0 \qquad \text{for all} \quad x = b_1 + vb_2 \neq 0 \quad (b_i \text{ in } \mathfrak{Q}).$$

A necessary condition is clearly that $\mathfrak{Q}$ be a quaternion division algebra. Then (3.48) implies that $\mathfrak{C}$ is a division algebra if and only if $\mathfrak{Q}$ is a division algebra and μ is not the norm $n(b)$ of an element b in $\mathfrak{Q}$; that is, $\mu \ (= \mu_3)$ is not represented by the quadratic form

$$X_1{}^2 + X_1 X_2 - \alpha X_2{}^2 - \mu_2 X_3{}^2 - \mu_2 X_3 X_4 + \alpha\mu_2 X_4{}^2.$$

For characteristic $\neq 2$, we may change the basis for $\mathfrak{Q}$ to see that $\mathfrak{C}$ is a division algebra if and only if $\mathfrak{Q}$ is a division algebra and $\mu (= \mu_3)$ is not represented by the quadratic form $X_1{}^2 - \mu_1 X_2{}^2 - \mu_2 X_3{}^2 + \mu_1\mu_2 X_4{}^2$. Quite trivially, there are no finite Cayley division algebras since, by Wedderburn's theorem on finite associative division rings (Artin [2], p. 37; Albert [24], p. 62), there are no finite quaternion division algebras.

6. THE WEDDERBURN PRINCIPAL THEOREM

A finite-dimensional algebra $\mathfrak{A}$ over F is called *separable* in case, for every extension K of F, the algebra $\mathfrak{A}_K$ is a direct sum of simple ideals. For alternative algebras this is equivalent to saying that $\mathfrak{A} = \mathfrak{S}_1 \oplus \cdots \oplus \mathfrak{S}_t$ is semisimple with the center $\mathfrak{C}_i$ of each simple component $\mathfrak{S}_i$ being a separable extension of F $(i = 1, \ldots, t)$.

The following theorem generalizes to alternative algebras the well-known *Wedderburn principal theorem* for associative algebras. We shall not repeat the portions of the proof which merely repeat the associative case (Albert [24], p. 47). We first prove three lemmas which will be used in the proof of

Theorem 3.18 (Wedderburn principal theorem for alternative algebras). *Let $\mathfrak{A}$ be a finite-dimensional alternative algebra over F with radical $\mathfrak{N}$. If $\mathfrak{A}/\mathfrak{N}$ is separable, then*

$$(3.49) \qquad \mathfrak{A} = \mathfrak{S} + \mathfrak{N} \qquad \text{(direct sum)},$$

where $\mathfrak{S}$ is a subalgebra of $\mathfrak{A}, \mathfrak{S} \cong \mathfrak{A}/\mathfrak{N}$.

Throughout the proof we shall write $[x]$ for the residue class $[x] = x + \mathfrak{N}$ of x modulo $\mathfrak{N}$, x in $\mathfrak{A}$.

Lemma 3.19. *Let $\mathfrak{A}$ be a finite-dimensional alternative algebra with 1 over F and with radical $\mathfrak{N}$. If $[u_1], \ldots, [u_t]$ are pairwise orthogonal idempotents in $\mathfrak{A}/\mathfrak{N}$, then they may be "lifted" to pairwise orthogonal idempotents $e_1, \ldots, e_t$ in $\mathfrak{A}$ satisfying $[e_i] = [u_i]$ $(i = 1, \ldots, t)$. If $[1] = [u_1] + \cdots + [u_t]$, then $1 = e_1 + \cdots + e_t$.*

Proof. If $[u]$ is idempotent in $\mathfrak{A}/\mathfrak{N}$, then any representative u of $[u]$ is not nilpotent. Since the subalgebra of $\mathfrak{A}$ generated by u is not nilpotent, $\mathfrak{A}$ contains an idempotent $e = \sum \alpha_i u^i$, α_i in F, and $[e] = \beta[u]$ for β in F. Since e is not in $\mathfrak{N}$, we have

$$[e] \neq 0 \qquad \text{and} \qquad [e] = [e]^2 = \beta^2[u]^2 = \beta^2[u] = \beta[e],$$

implying $\beta = 1$, $[e] = [u]$. This is the case $t = 1$. We assume that $e_1, \ldots, e_{t-1}$ are pairwise orthogonal idempotents of $\mathfrak{A}$ satisfying $[e_i] = [u_i]$ $(i = 1, \ldots, t-1)$. Then $e = e_1 + \cdots + e_{t-1}$ is idempotent, and in the Peirce decomposition (3.14) of $\mathfrak{A}$ relative to e we have $\mathfrak{A}_{00} = f\mathfrak{A}f$ where $f = 1 - e$. Correspondingly we have, in the Peirce decomposition of $\mathfrak{A}/\mathfrak{N}$ relative to $[e]$, $(\mathfrak{A}/\mathfrak{N})_{00} = [f](\mathfrak{A}/\mathfrak{N})[f]$. Since $[u_t]$ is in $(\mathfrak{A}/\mathfrak{N})_{00} = [f](\mathfrak{A}/\mathfrak{N})[f]$, we may take u_t in $f\mathfrak{A}f = \mathfrak{A}_{00}$. By Corollary 3.8, the radical of $\mathfrak{A}_{00}$ is $\mathfrak{N} \cap \mathfrak{A}_{00} = f\mathfrak{N}f$. Hence

$$[u_t] \in [f](\mathfrak{A}/\mathfrak{N})[f] = (\mathfrak{A}_{00} + \mathfrak{N})/\mathfrak{N} \cong \mathfrak{A}_{00}/(\mathfrak{N} \cap \mathfrak{A}_{00})$$

implies that we may apply the case $t = 1$ to $\mathfrak{A}_{00}$ to see that the residue class $u_t + f\mathfrak{N}f$ in $\mathfrak{A}_{00}/(\mathfrak{N} \cap \mathfrak{A}_{00})$ may be lifted to an idempotent e_t in $\mathfrak{A}_{00}$ satisfying $e_t + f\mathfrak{N}f = u_t + f\mathfrak{N}f$, and therefore $[e_t] = [u_t]$. Since e_t is in $\mathfrak{A}_{00}$, we have $e_1, \ldots, e_t$ pairwise orthogonal. If $[1] = [u_1] + \cdots + [u_t]$, then $1 - (e_1 + \cdots + e_t)$ is in $\mathfrak{N}$. But $1 - (e_1 + \cdots + e_t)$ is either 0 or idempotent. Hence $1 = e_1 + \cdots + e_t$.

Lemma 3.20. *Let $\mathfrak{A}$ be a finite-dimensional alternative algebra with 1 over F. Let $\mathfrak{N}$ be the radical of $\mathfrak{A}$, and assume $\mathfrak{N}^2 = 0$. If F_t is a total matrix subalgebra of $\mathfrak{A}/\mathfrak{N}$ which contains the identity element $[1]$ of $\mathfrak{A}/\mathfrak{N}$, then $\mathfrak{A}$ contains a total matrix algebra $\mathfrak{M}$ of degree t with identity element 1, and F_t is the image of $\mathfrak{M}$ under the natural homomorphism $x \to [x]$ of $\mathfrak{A}$ onto $\mathfrak{A}/\mathfrak{N}$.*

Proof. Let F_t have basal elements $[u_{ij}]$ $(i, j = 1, \ldots, t)$ with the familiar multiplication table

$$[u_{ij}][u_{kl}] = \delta_{jk}[u_{il}].$$

We wish to lift these elements to (clearly linearly independent) elements e_{ij} of $\mathfrak{A}$ with the same multiplication table. By Lemma 3.19 there exist pairwise orthogonal idempotents $e_{11}, \ldots, e_{tt}$ in $\mathfrak{A}$ such that $[e_{ii}] = [u_{ii}]$ $(i = 1, \ldots, t)$ and $1 = e_{11} + \cdots + e_{tt}$. Let (3.15) be the corresponding Peirce decomposition of $\mathfrak{A}$. We select representatives u_{i1} (resp. u_{1j}) of $[u_{i1}]$ (resp. $[u_{1j}]$) as follows. We may take $u_{11} = e_{11}$. Since

$$[u_{i1}] = [e_{ii}][u_{i1}][e_{11}],$$

we may also take u_{i1} in $\mathfrak{A}_{i1}$ $(i = 2, \ldots, t)$, and similarly take u_{1j} in $\mathfrak{A}_{1j}$ $(j = 2, \ldots, t)$. It follows that

$$u_{1j}u_{j1} = e_{11} + z_j \qquad (j = 2, \ldots, t)$$

where z_j is in $\mathfrak{N} \cap \mathfrak{A}_{11}$ by (3.17). Write

$$e_{1j} = (e_{11} - z_j)u_{1j}, \quad e_{i1} = u_{i1} \qquad (i, j = 2, \ldots, t).$$

Then e_{1j} is in $\mathfrak{A}_{11}\mathfrak{A}_{1j} \subseteq \mathfrak{A}_{1j}$ and e_{i1} is in $\mathfrak{A}_{i1}$. Also

$$(3.50) \qquad\qquad e_{1j}e_{j1} = e_{11} \qquad (j = 1, \ldots, t).$$

For $e_{11}^2 = e_{11}$ and, if $j \neq 1$, we have

$$e_{1j}e_{j1} = [(e_{11} - z_j)u_{1j}]u_{j1} = (e_{11} - z_j)(u_{1j}u_{j1})$$
$$= (e_{11} - z_j)(e_{11} + z_j) = e_{11}$$

by (3.22), since z_j is in $\mathfrak{A}_{11}$ and $z_j{}^2$ is in $\mathfrak{N}^2 = 0$. Define

(3.51) $\qquad\qquad e_{ij} = e_{i1}e_{1j} \qquad (i \neq j; \quad i, j = 2, \ldots, t)$

in $\mathfrak{A}_{i1}\mathfrak{A}_{1j} \subseteq \mathfrak{A}_{ij}$. We have

$$[e_{ij}] = [e_{i1}][e_{1j}] = [u_{i1}]([e_{11} - z_j][u_{1j}])$$
$$= [u_{i1}]([u_{11}][u_{1j}]) = [u_{ij}].$$

Now (3.51) holds for $i, j = 1, \ldots, t$. For $i = 1$ or $j = 1$ this is trivial. Also

$$(e_{i1}e_{1i})^2 = e_{i1}[(e_{1i}e_{i1})e_{1i}] = e_{i1}e_{1i}$$

by Artin's Theorem 3.1; since $[e_{i1}e_{1i}] = [u_{ii}] = [e_{ii}]$ we may apply the last sentence of Lemma 3.19 to $\mathfrak{A}_{ii}$ to obtain $e_{ii} = e_{i1}e_{1i}$. It follows from (3.19) and (3.20) that

(3.52) $\qquad\qquad e_{ij}e_{kl} = 0 \qquad \text{for} \quad j \neq k.$

To see that

(3.53) $\qquad\qquad e_{ij}e_{jk} = e_{ik} \qquad (i, j, k = 1, \ldots, t),$

we linearize the Moufang identity (3.6) to obtain

$$(xy)(az) + (zy)(az) = x[(ya)z] + z[(ya)x]$$

for all x, y, a, z in $\mathfrak{A}$. Putting $x = e_{i1}$, $y = e_{1j}$, $a = e_{j1}$, $z = e_{1k}$ in this identity we have

$$e_{ij}e_{jk} = (e_{i1}e_{1j})(e_{j1}e_{1k})$$
$$= -(e_{1k}e_{1j})(e_{j1}e_{i1}) + e_{i1}[(e_{1j}e_{j1})e_{1k}] + e_{1k}[(e_{1j}e_{j1})e_{i1}]$$
$$= -\delta_{k1}\delta_{1i}e_{11} + e_{i1}e_{1k} + \delta_{k1}\delta_{1i}e_{11}$$
$$= e_{ik}$$

by (3.51), (3.52), and (3.50). It follows from (3.52) and (3.53) that the subalgebra $\mathfrak{M}$ of $\mathfrak{A}$ with basis $\{e_{ij}\}$ $(i, j = 1, \ldots, t)$ is a total matrix algebra, completing the proof.

Lemma 3.21. *Let* $\mathfrak{A}$ *be a finite-dimensional alternative algebra with* 1 *over* F. *Let* $\mathfrak{N}$ *be the radical of* $\mathfrak{A}$, *and assume* $\mathfrak{N}^2 = 0$. *If* $\mathfrak{A}/\mathfrak{N}$ *is a split Cayley algebra, then* $\mathfrak{A}$ *contains a subalgebra* $\mathfrak{C} \cong \mathfrak{A}/\mathfrak{N}$.

Proof. We may take $\mathfrak{A}/\mathfrak{N}$ in the form $\mathfrak{A}/\mathfrak{N} = F_2 + [w]F_2$ of Lemma 3.16, $[w]^2 = [1]$, multiplication given by (3.38) with the obvious changes in notation. By Lemma 3.20, $\mathfrak{A}$ contains a total matrix algebra $\mathfrak{Q} \cong F_2$ such that $\mathfrak{Q}$ contains 1 and the matric basis $\{e_{ij}\}$ of $\mathfrak{Q}$ yields the matric basis $\{[e_{ij}]\}$ of F_2. Note that $\overline{[b]} = [\bar{b}]$ for b in $\mathfrak{Q}$. We have already observed that

$$\bar{e}_{11} = e_{22}, \qquad \bar{e}_{12} = -e_{12}, \qquad \bar{e}_{21} = -e_{21}, \qquad \bar{e}_{22} = e_{11}.$$

In order to prove the lemma, it is sufficient to show the existence of $v \notin \mathfrak{Q}$ satisfying $v^2 = 1$ and $bv = v\bar{b}$ for all b in $\mathfrak{Q}$.
    Write

$$[f_{ij}] = [w][e_{jj}] \qquad \text{for} \quad i \neq j \quad (i, j = 1, 2).$$

Using the Peirce decomposition of $\mathfrak{A}$ relative to $e_1 = e_{11}$, $e_2 = e_{22}$, we may take f_{ij} in $\mathfrak{A}_{ij}$ $(i \neq j)$, for

$$[e_{ii}]([f_{ij}][e_{jj}]) = [e_{ii}]([w][e_{jj}]^2) = [w][\bar{e}_{ii}e_{jj}] = [w][e_{jj}] = [f_{ij}].$$

Now

$$[e_{ji}][f_{ij}] = [e_{ji}]([w][e_{ij}]) = -[w]([e_{ji}][e_{jj}]) = 0,$$

implying that

$$e_{ji}f_{ij} = c_j, \qquad c_j \quad \text{in} \quad \mathfrak{N} \cap \mathfrak{A}_{jj} \qquad (i \neq j; \;\; i, j = 1, 2).$$

Write

$$h_{ij} = f_{ij} - e_{ij}c_j.$$

Then h_{ij} is in $\mathfrak{A}_{ij}$, $[h_{ij}] = [f_{ij}]$, and

$$e_{ji}h_{ij} = h_{ij}e_{ji} = 0 \qquad (i \neq j; \;\; i, j = 1, 2).$$

For

$$e_{ji}h_{ij} = c_j - e_{ji}(e_{ij}c_j) = c_j - (e_{ji}e_{ij})c_j = 0$$

by (3.22). Also

$$e_{ij}c_j = e_{ij}(e_{ji}f_{ij}) = (e_{ij}e_{ji})f_{ij} - (e_{ij}, e_{ji}, f_{ij})$$

$$= f_{ij} + (f_{ij}, e_{ji}, e_{ij}) = f_{ij} + (f_{ij}e_{ji})e_{ij} - f_{ij} = (f_{ij}e_{ji})e_{ij},$$

so that

$$h_{ij}e_{ji} = f_{ij}e_{ji} - [(f_{ij}e_{ji})e_{ij}]e_{ji} = 0$$

by (3.22). Now

$$[h_{ij}][h_{ji}] = [f_{ij}][f_{ji}] = [e_{ii}][\bar{e}_{jj}] = [e_{ii}]$$

implies that

$$h_{ij}h_{ji} = e_{ii} + a_i, \qquad a_i \text{ in } \mathfrak{N} \cap \mathfrak{A}_{ii} \qquad (i \neq j; \quad i, j = 1, 2).$$

Then $a_i^2 = 0$ since $\mathfrak{N}^2 = 0$, and

$$(e_{ii} - a_i)(e_{ii} + a_i) = e_{ii} = (e_{ii} + a_i)(e_{ii} - a_i) \qquad (i = 1, 2).$$

Write

$$p_{12} = (e_{11} - a_1)h_{12}, \qquad p_{21} = h_{21}.$$

Then p_{ij} is in $\mathfrak{A}_{ij}$, $[p_{ij}] = [f_{ij}]$, and we shall prove

$$p_{ij}p_{ji} = e_{ii} \qquad (i \neq j; \quad i, j = 1, 2).$$

For

$$\begin{aligned}
p_{12}p_{21} &= [(e_{11} - a_1)h_{12}]h_{21} = (e_{11} - a_1)(h_{12}h_{21}) \\
&= (e_{11} - a_1)(e_{11} + a_1) = e_{11}.
\end{aligned}$$

But

$$\begin{aligned}
a_i h_{ij} &= (h_{ij}h_{ji} - e_{ii})h_{ij} = h_{ij}(h_{ji}h_{ij}) - h_{ij} \\
&= h_{ij}(e_{jj} + a_j) - h_{ij} = h_{ij}a_j,
\end{aligned}$$

so that

$$p_{12} = h_{12} - a_1 h_{12} = h_{12} - h_{12}a_2 = h_{12}(e_{22} - a_2)$$

and

$$p_{21}p_{12} = h_{21}[h_{12}(e_{22} - a_2)] = (h_{21}h_{12})(e_{22} - a_2) = e_{22}.$$

Also (3.22) implies that

$$e_{ij}p_{ji} = p_{ji}e_{ij} = 0 \qquad (i \neq j; \quad i, j = 1, 2).$$

Finally write

$$v = p_{12} + p_{21}.$$

Then $[v] = [f_{12}] + [f_{21}] = [w]$, implying $v \notin Q$. Also

$$v^2 = (p_{12} + p_{21})^2 = e_{11} + e_{22} = 1$$

by (3.20). Writing

$$b = \alpha e_{11} + \beta e_{12} + \gamma e_{21} + \delta e_{22},$$

we have

$$\bar{b} = \delta e_{11} - \beta e_{12} - \gamma e_{21} + \alpha e_{22},$$

$$bv = \alpha p_{12} + \beta e_{12} p_{12} + \gamma e_{21} p_{21} + \delta p_{21}$$

$$= \alpha p_{12} - \beta p_{12} e_{12} - \gamma p_{21} e_{21} + \delta p_{21} = v\bar{b}$$

by (3.21), completing the proof of the lemma.

Proof of Theorem 3.18.   It suffices to prove the existence in $\mathfrak{A}$ of $\mathfrak{S} \cong \mathfrak{A}/\mathfrak{N}$. Since the theorem is trivial unless $\mathfrak{N} \neq 0$, and since $\mathfrak{N}$ is solvable, we have proper inclusions in the derived series

$$\mathfrak{N} = \mathfrak{N}^{(1)} \supset \mathfrak{N}^{(2)} \supset \cdots \supset \mathfrak{N}^{(r)} = 0.$$

Also $\mathfrak{N}^2 \, (= \mathfrak{N}^{(2)})$ is an ideal of $\mathfrak{A}$. For a in $\mathfrak{A}$ and x, y in $\mathfrak{N}$ imply

$$a(xy) = (ax)y - (a, x, y) = (ax)y - (y, a, x)$$

$$= (ax)y - (ya)x + y(ax)$$

is in $\mathfrak{N}^2$ since $\mathfrak{N}$ is an ideal; hence $\mathfrak{N}^2$ is a left ideal of $\mathfrak{A}$. Reciprocally, $\mathfrak{N}^2$ is a right ideal of $\mathfrak{A}$. The same inductive argument based on the dimension of $\mathfrak{A}$ which is used for associative algebras suffices to reduce the proof of the theorem to the case $\mathfrak{N}^2 = 0$.

We have shown in Lemmas 3.20 and 3.21 that the theorem is true for $\mathfrak{A}$ containing 1, with $\mathfrak{N}^2 = 0$, and $\mathfrak{A}/\mathfrak{N}$ either a total matrix algebra or a split Cayley algebra. We now reduce the case $\mathfrak{N}^2 = 0$ to one of these two situations. If $\mathfrak{A}$ has no identity element, adjoin 1 to $\mathfrak{A}$ to obtain $\mathfrak{A}_1 = F1 + \mathfrak{A}$. Then $\mathfrak{A}_1$ is alternative. Since $\mathfrak{N}$ is a nilpotent ideal of $\mathfrak{A}_1$, we have $\mathfrak{N} \subseteq \mathfrak{N}_1$, the radical of $\mathfrak{A}_1$. Conversely, a_1 in $\mathfrak{N}_1$ implies $a_1 = \alpha 1 + a$, α in F, a in $\mathfrak{A}$, and $0 = a_1{}^t = \alpha^t 1 + b$, b in $\mathfrak{A}$. Hence $\alpha = 0$, $\mathfrak{N}_1 \subseteq \mathfrak{A}$. Hence $\mathfrak{N}_1$ is a nilpotent ideal of $\mathfrak{A}$, $\mathfrak{N}_1 \subseteq \mathfrak{N}$. If we can obtain a

Wedderburn decomposition $\mathfrak{A}_1 = \mathfrak{S}_1 + \mathfrak{N}$ of $\mathfrak{A}_1$, we have $\mathfrak{A} = \mathfrak{S} + \mathfrak{N}$ where $\mathfrak{S} = \mathfrak{S}_1 \cap \mathfrak{A} \cong \mathfrak{A}/\mathfrak{N}$ as desired.

We assume henceforth that $\mathfrak{A}$ contains 1, and now reduce the proof of the theorem to the case where $\mathfrak{A}/\mathfrak{N}$ is simple. Let $\mathfrak{A}/\mathfrak{N} = \mathfrak{B}_1 \oplus \cdots \oplus \mathfrak{B}_t$ for simple ideals $\mathfrak{B}_i$. By Lemma 3.19 we can lift the identity elements of the $\mathfrak{B}_i$ to pairwise orthogonal idempotents e_i in $\mathfrak{A}$ satisfying $1 = e_1 + \cdots + e_t$. In the corresponding Peirce decomposition (3.15) the radical of $\mathfrak{A}_{ii}$ is $\mathfrak{N}_i = \mathfrak{N} \cap \mathfrak{A}_{ii}$ ($i = 1, \ldots, t$) by Corollary 3.8. Now $\mathfrak{A}_{ii}/\mathfrak{N}_i \cong \mathfrak{B}_i$. If we have Wedderburn decompositions $\mathfrak{A}_{ii} = \mathfrak{S}_i + \mathfrak{N}_i$, $\mathfrak{S}_i \cong \mathfrak{B}_i$, then (since the subalgebras $\mathfrak{A}_{ii}$ of $\mathfrak{A}$ are pairwise orthogonal) the sum of the subalgebras $\mathfrak{S}_i$ of $\mathfrak{A}$ is the direct sum

$$\mathfrak{S} = \mathfrak{S}_1 \oplus \cdots \oplus \mathfrak{S}_t \cong \mathfrak{A}/\mathfrak{N}.$$

Let K be the algebraic closure of F. Since $\mathfrak{A}/\mathfrak{N}$ is separable, $(\mathfrak{A}/\mathfrak{N})_K = \mathfrak{A}_K/\mathfrak{N}_K$ is a direct sum of total matrix algebras K_n and split Cayley algebras. Then $\mathfrak{N}_K$ is the radical of $\mathfrak{A}_K$ (since it is a nilpotent ideal containing the radical). By the result in the preceding paragraph, we may assume that $(\mathfrak{A}/\mathfrak{N})_K$ has only one simple component. Then, by Lemma 3.20 or 3.21, $\mathfrak{A}_K$ contains a subalgebra $\mathfrak{B} \cong (\mathfrak{A}/\mathfrak{N})_K$. The remaining steps are those of the associative proof, since no form of associativity is used there.

Theorem 3.18 is the analogue for alternative algebras of Levi's theorem for Lie algebras of characteristic 0 (Jacobson [25], p. 91). Levi's theorem is proved, not by the structure lifting which involves an examination of the individual split simple algebras as above, but by use of the second Whitehead lemma for Lie algebras (Jacobson [25], p. 89). The associative case of Theorem 3.18 for characteristic 0 has a similar proof (Hochschild [1]). The second Whitehead lemma for alternative algebras (of arbitrary characteristic) is valid, since it is equivalent to the case $\mathfrak{N}^2 = 0$ of Theorem 3.18. It would be interesting to have a proof (if only for characteristic 0) which does not involve an examination of individual cases. In order to state the second Whitehead lemma for alternative algebras, it is necessary to introduce alternative bimodules, which we shall do in the fashion indicated at the end of Chapter II.

Let $\mathfrak{A}$ be an alternative algebra over an arbitrary field F, and $\mathfrak{M}$ be a vector space over F. Then $\mathfrak{M}$ is an *alternative bimodule* for $\mathfrak{A}$ in case there are two bilinear compositions ma, am in $\mathfrak{M}$ (a in $\mathfrak{A}$, m in $\mathfrak{M}$) satisfying

(3.54) $(a, a, m) = (m, a, a) = 0$ for all a in $\mathfrak{A}$, m in $\mathfrak{M}$,

and

(3.55) $(a, m, b) = -(m, a, b)$, $(b, a, m) = -(b, m, a)$

for all a, b in $\mathfrak{A}$, m in $\mathfrak{M}$, where the associator is defined as in Chapter II, except that one argument is in $\mathfrak{M}$. If F has characteristic $\neq 2$, then (3.54) and (3.55) together are equivalent to

$$(a, m, b) = -(m, a, b) = (b, a, m) = -(b, m, a)$$

for all a, b in $\mathfrak{A}$, m in $\mathfrak{M}$. The vector space direct sum $\mathfrak{A} + \mathfrak{M}$ is made into an algebra, called the *split null extension* or *semidirect sum*, by defining multiplication by

$$(a + m_1)(b + m_2) = ab + (m_1 b + a m_2)$$

for all a, b in $\mathfrak{A}$, m_i in $\mathfrak{M}$. Having $\mathfrak{M}$ an alternative bimodule for $\mathfrak{A}$ is equivalent to having $\mathfrak{A} + \mathfrak{M}$ an alternative algebra, since the alternative law $(x, x, y) = (y, x, x) = 0$ in $\mathfrak{A} + \mathfrak{M}$, $x = a + m_1$, $y = b + m_2$, is equivalent to the alternative law in $\mathfrak{A}$, (3.54), and (3.55). Clearly $\mathfrak{M}$ is an ideal of $\mathfrak{A} + \mathfrak{M}$, and $\mathfrak{M}^2 = 0$.

Let $\mathfrak{M}$ be an alternative bimodule for the alternative algebra $\mathfrak{A}$, and consider the linear operators S_a, T_a (a in $\mathfrak{A}$) defined by

$$m \to m S_a = ma, \qquad m \to m T_a = am \qquad \text{for all} \quad m \text{ in } \mathfrak{M}.$$

By (3.54) and (3.55) we have

(3.56) $T_{a^2} = T_a{}^2$, $S_a{}^2 = S_{a^2}$ for all a in $\mathfrak{A}$

and

(3.57) $[T_a, S_b] = -S_a S_b + S_{ab}$, $T_{ba} - T_a T_b = -[T_b, S_a]$

for all a, b in $\mathfrak{A}$. A pair (S, T) of linear mappings $a \to S_a$, $a \to T_a$ of an alternative algebra $\mathfrak{A}$ over F into the algebra $\mathfrak{E}(\mathfrak{M})$ of all linear operators on some vector space $\mathfrak{M}$ over F is called a *(bi)representation* of $\mathfrak{A}$ acting in $\mathfrak{M}$ in case (3.56) and (3.57) are satisfied. Thus every alternative bimodule for $\mathfrak{A}$ is associated with a representation. The converse is also true. For, if (S, T) is a representation of $\mathfrak{A}$ acting in $\mathfrak{M}$, then $\mathfrak{M}$ is an alternative bimodule for $\mathfrak{A}$ relative to the two compositions

$$ma = mS_a, \qquad am = mT_a \qquad (a \text{ in } \mathfrak{A}, \quad m \text{ in } \mathfrak{M}).$$

Every alternative algebra $\mathfrak{A}$ over F has a representation; namely, the *regular representation* (R, L) where R_a and L_a are the right and left multiplications of $\mathfrak{A}$ corresponding to a in $\mathfrak{A}$. In this case $\mathfrak{A}$ itself is the alternative bimodule. More generally, if $\mathfrak{A}$ is any subalgebra of an alternative algebra $\mathfrak{B}$ over F, and if $\mathfrak{M}$ is an ideal of $\mathfrak{B}$, then the regular representation of $\mathfrak{B}$ induces a representation of $\mathfrak{A}$ acting in $\mathfrak{M}$. If (S, T) is any representation of an alternative algebra $\mathfrak{A}$ acting in $\mathfrak{M}$, then (S, T) is induced by the regular representation of the split null extension $\mathfrak{A} + \mathfrak{M}$.

We shall defer statement of the first Whitehead lemma for alternative algebras until later in this chapter because derivations of alternative algebras are involved, with a restriction imposed upon the characteristic of F. However, for an arbitrary field F we have

Proposition 3.22 (Second Whitehead lemma for alternative algebras). *Let $\mathfrak{A}$ be a finite-dimensional separable alternative algebra, and let $\mathfrak{M}$ be a finite-dimensional alternative bimodule for $\mathfrak{A}$. Let f be a bilinear mapping of $\mathfrak{A}$ into $\mathfrak{M}$ satisfying*

$$(3.58) \qquad F(a, a, b) = F(b, a, a) = 0 \qquad \text{for all} \quad a, b \text{ in } \mathfrak{A},$$

where

$$F(a, b, c) = f(a, b)c + f(ab, c) - af(b, c) - f(a, bc).$$

Then there exists a linear mapping g of $\mathfrak{A}$ into $\mathfrak{M}$ such that

$$(3.59) \quad f(a, b) = ag(b) + g(a)b - g(ab) \qquad \text{for all} \quad a, b \text{ in } \mathfrak{A}.$$

Proof. The second Whitehead lemma for alternative algebras is equivalent to the case $\mathfrak{N}^2 = 0$ of the Wedderburn principal theorem which we have already proved.

We first give a proof of Proposition 3.22. Let $\mathfrak{B} = \mathfrak{A} + \mathfrak{M}$ be the vector space direct sum of $\mathfrak{A}$ and $\mathfrak{M}$. Define a multiplication $x * y$ in $\mathfrak{B}$ by

$$(a + m_1) * (b + m_2) = ab + f(a, b) + am_2 + m_1 b$$

for a, b in $\mathfrak{A}$, m_i in $\mathfrak{M}$. Writing

$$(x, y, z)_* = (x * y) * z - x * (y * z) \qquad \text{for} \quad x, y, z \text{ in } \mathfrak{B},$$

we see, since $\mathfrak{M}^2 = 0$, that all such associators with more than one argument in $\mathfrak{M}$ are 0, while for a, b in $\mathfrak{A}$, m in $\mathfrak{M}$, we have

$$(a, b, m)_* = (a, b, m), \quad (a, m, b)_* = (a, m, b), \quad (m, a, b)_* = (m, a, b).$$

Also

$$(a, b, c)_* = (a, b, c) + F(a, b, c) \qquad \text{for} \quad a, b, c \text{ in } \mathfrak{A}.$$

Hence the alternative law in $\mathfrak{B}$ is implied by the alternative law in $\mathfrak{A}$, (3.54), (3.55), and (3.58).

Now $\mathfrak{B}/\mathfrak{M} \cong \mathfrak{A}$ is separable, so that $\mathfrak{M}$ is the radical of $\mathfrak{B}$. The case $\mathfrak{M}^2 = 0$ of the Wedderburn principal theorem implies that

$$\mathfrak{B} = \mathfrak{S} + \mathfrak{M} \qquad \text{(direct sum)}$$

where $\mathfrak{S}$ is a subalgebra of $\mathfrak{B}$. We may write any element a of $\mathfrak{A} \subseteq \mathfrak{B} = \mathfrak{S} + \mathfrak{M}$ uniquely in the form

$$a = s(a) + g(a), \qquad s(a) \text{ in } \mathfrak{S}, \quad g(a) \text{ in } \mathfrak{M}.$$

The mapping $a \to g(a)$ is obviously linear. Now $\mathfrak{S}$ contains

$$\begin{aligned}
s(a) * s(b) &= (a - g(a)) * (b - g(b)) \\
&= ab + f(a, b) - ag(b) - g(a)b \\
&= s(ab) + g(ab) + f(a, b) - ag(b) - g(a)b
\end{aligned}$$

for all a, b in $\mathfrak{A}$. Since $\mathfrak{B} = \mathfrak{S} + \mathfrak{M}$ is a direct sum, we have (3.59) as desired.

Conversely, assuming Proposition 3.22, we can prove the case $\mathfrak{N}^2 = 0$ of Theorem 3.18. For, if $\mathfrak{A}$ is an alternative algebra with radical $\mathfrak{N}$, there is a subspace $\mathfrak{T}$ of $\mathfrak{A}$ such that

$$\mathfrak{A} = \mathfrak{T} + \mathfrak{N} \qquad \text{(direct sum)}.$$

Thus there is a vector space isomorphism $a \to a^*$ between $\mathfrak{A}/\mathfrak{N}$ and $\mathfrak{T}$ (a in $\mathfrak{A}/\mathfrak{N}$, a^* in $\mathfrak{T}$) such that $(ab)^* - a^*b^*$ is in $\mathfrak{N}$ for all a, b in $\mathfrak{A}/\mathfrak{N}$. Write

$$f(a, b) = (ab)^* - a^*b^*.$$

Since $\mathfrak{N}^2 = 0$, $\mathfrak{N}$ is an alternative bimodule for $\mathfrak{A}/\mathfrak{N}$ when we define

$$ma = ma^*, \quad am = a^*m \qquad \text{for} \quad a \text{ in } \mathfrak{A}/\mathfrak{N}, \quad m \text{ in } \mathfrak{N}.$$

Also f is a bilinear mapping of $\mathfrak{A}/\mathfrak{N}$ into $\mathfrak{N}$ satisfying (3.58). For compute

$$\begin{aligned}
F(a, b, c) &= (ab)^*c^* - (a^*b^*)c^* + ((ab)c)^* - (ab)^*c^* \\
&\quad - a^*(bc)^* + a^*(b^*c^*) - (a(bc))^* + a^*(bc)^* \\
&= (a, b, c)^* - (a^*, b^*, c^*).
\end{aligned}$$

Then (3.58) follows from the alternative law in $\mathfrak{A}$. Since $\mathfrak{A}/\mathfrak{N}$ is separable, the second Whitehead lemma guarantees the existence of a linear mapping $a \to g(a)$ of $\mathfrak{A}/\mathfrak{N}$ into $\mathfrak{N}$ satisfying

$$(ab)^* - a^*b^* = a^*g(b) + g(a)b^* - g(ab) \qquad \text{for all} \quad a, b \text{ in } \mathfrak{A}/\mathfrak{N}.$$

Setting $a' = a^* + g(a)$ in $\mathfrak{A}$, we have

$$a'b' = a^*b^* + a^*g(b) + g(a)b^* = (ab)^* + g(ab) = (ab)',$$

so that $a \to a'$ is an isomorphism of the algebra $\mathfrak{A}/\mathfrak{N}$ into $\mathfrak{A}$, completing the proof.

7. NORM FORMS

We assume that the characteristic of F is $\neq 2$ to develop now some properties of Cayley algebras which depend on the nondegeneracy of the norm form $n(x)$. All of the algebras $\mathfrak{A} = \mathfrak{B} + v\mathfrak{B}$ constructed from $F1$ by iteration of the Cayley–Dickson process have nondegenerate norm forms, as may be seen inductively as follows.

Assume that the norm on $\mathfrak{B}$ is a nondegenerate quadratic form; that is, the associated symmetric bilinear form

$$(3.60) \qquad (a, b) = \tfrac{1}{2}[n(a + b) - n(a) - n(b)] = \tfrac{1}{2}t(a\bar{b})$$

is nondegenerate [if $(a, b) = 0$ for all b in $\mathfrak{B}$, then $a = 0$]. Then the norm $n(x) = n(b_1) - \mu n(b_2)$ on $\mathfrak{A}$ given by (3.39), $x = b_1 + vb_2$, is nondegenerate. For $y = b_3 + vb_4$ implies that

$$(x, y) = \tfrac{1}{2}[n(x + y) - n(x) - n(y)]$$
$$= \tfrac{1}{2}[n(b_1 + b_3) - \mu n(b_2 + b_4) - n(b_1) + \mu n(b_2) - n(b_3) + \mu n(b_4)]$$
$$= (b_1, b_3) - \mu(b_2, b_4).$$

Hence $(x, y) = 0$ for all $y = b_3 + vb_4$ implies $(b_1, b_3) = \mu(b_2, b_4)$ for all b_3, b_4 in $\mathfrak{B}$. Then $b_4 = 0$ implies $(b_1, b_3) = 0$ for all b_3 in $\mathfrak{B}$, or $b_1 = 0$ since $n(b)$ is nondegenerate on $\mathfrak{B}$; similarly $b_3 = 0$ implies $(b_2, b_4) = 0$ (since $\mu \neq 0$) for all b_4 in $\mathfrak{B}$, or $b_2 = 0$. That is, $x = 0$; $n(x)$ is nondegenerate on $\mathfrak{A}$. (At the first step in the induction, α^2 is clearly nondegenerate on $F1$.)

Theorem 3.23 (Jacobson). *Two Cayley algebras $\mathfrak{C}$ and $\mathfrak{C}'$ of characteristic $\neq 2$ are isomorphic if and only if their corresponding norm forms $n(x)$ and $n'(x')$ are equivalent (that is, there is a linear mapping $x \to xH$ of $\mathfrak{C}$ into $\mathfrak{C}'$ such that*

(3.61) $$n'(xH) = n(x) \qquad \text{for all} \quad x \text{ in } \mathfrak{C};$$

H is necessarily (1-1) since $n(x)$ is nondegenerate).

Proof. Suppose $\mathfrak{C}$ and $\mathfrak{C}'$ are isomorphic, the isomorphism being H. Then (3.35) implies $(xH)^2 - t(x)(xH) + n(x)1' = 0$ where $1' = 1H$ is the unity element of $\mathfrak{C}'$. But also $(xH)^2 - t'(xH)(xH) + n'(xH)1' = 0$. Hence

$$[t'(xH) - t(x)](xH) + [n(x) - n'(xH)]1' = 0.$$

If $x \notin F1$, then $xH \notin F1'$ and $n(x) = n'(xH)$. On the other hand, $n(\alpha 1) = \alpha^2 = n'(\alpha 1')$, and we have (3.61) for all x in $\mathfrak{C}$.

For the converse we need to establish the fact that, if $\mathfrak{B}$ is a proper subalgebra of a Cayley algebra $\mathfrak{C}$, if $\mathfrak{B}$ contains the unity element 1 of $\mathfrak{C}$, and if (relative to the nondegenerate symmetric bilinear form (x, y) defined on $\mathfrak{C}$ by (3.60)) $\mathfrak{B}$ is a nonisotropic subspace of $\mathfrak{C}$ (that is, $\mathfrak{B} \cap \mathfrak{B}^\perp = 0$), then there is a subalgebra $\mathfrak{A} = \mathfrak{B} + v\mathfrak{B}$ obtained by the

Cayley–Dickson process. For the involution $x \to \bar{x}$ on $\mathfrak{C}$ induces an involution on $\mathfrak{B}$, since $\bar{b} = t(b)1 - b$ is in $\mathfrak{B}$ for all b in $\mathfrak{B}$. Also $\mathfrak{B}$ non-isotropic implies $\mathfrak{C} = \mathfrak{B} \perp \mathfrak{B}^{\perp}$ with $\mathfrak{B}^{\perp}$ nonisotropic (Jacobson [24], p. 151; Artin [2], p. 117). Hence there is a nonisotropic vector v in $\mathfrak{B}^{\perp}$, $n(v) = -\mu \neq 0$. Since $t(v) = t(v\bar{1}) = 2(v,1) = 0$, we have

$$v^2 = \mu 1 \qquad (\mu \neq 0 \text{ in } F).$$

Now

(3.62) $\qquad t((xy)z) = t(x(yz)) \qquad$ for all $\quad x, y, z$ in $\mathfrak{C}$.

For $(x, y, z) = -(z, y, x) = (\bar{z}, \bar{y}, \bar{x})$ implies $(xy)z + \bar{z}(\bar{y}\bar{x}) = x(yz) + (\bar{z}\bar{y})\bar{x}$. Hence $v\mathfrak{B} \subseteq \mathfrak{B}^{\perp}$ since (3.62) implies

$$(va, b) = \tfrac{1}{2}t((va)\bar{b}) = \tfrac{1}{2}t(v(a\bar{b})) = (v, b\bar{a}) = 0 \qquad \text{for all } a, b \text{ in } \mathfrak{B}.$$

Hence $\mathfrak{B} \perp v\mathfrak{B}$. Also $v\mathfrak{B}$ has the same dimension as $\mathfrak{B}$ since $b \to vb$ is (1-1). [Suppose $vb = 0$; then $v(vb) = v^2 b = \mu b = 0$, implying $b = 0$.] Now $t(v) = 0$ implies $\bar{v} = -v$; hence v in $\mathfrak{B}^{\perp}$ implies

$$0 = 2(v, b) = t(v\bar{b}) = v\bar{b} + b\bar{v} = v\bar{b} - bv,$$

or

$$bv = v\bar{b} \qquad \text{for all} \quad b \text{ in } \mathfrak{B}.$$

We have seen before that $v^2 = \mu 1$ and (3.40) imply (3.38). Hence $\mathfrak{A} = \mathfrak{B} \perp v\mathfrak{B}$ is the subalgebra specified. We have seen that, if $\mathfrak{B}$ is non-isotropic, then so is $\mathfrak{A}$.

Now let $\mathfrak{C}$ and $\mathfrak{C}'$ have equivalent norm forms $n(x)$ and $n'(x')$. Let $\mathfrak{B}$ (and $\mathfrak{B}'$) be as above. If $\mathfrak{B}$ and $\mathfrak{B}'$ are isomorphic under H_0, then the restrictions of $n(x)$ and $n'(x')$ to $\mathfrak{B}$ and $\mathfrak{B}'$ are equivalent. Then by Witt's theorem (Jacobson [24], p. 162; Artin [2], p. 121), since $n(x)$ and $n'(x')$ are equivalent, the restrictions of $n(x)$ and $n'(x')$ to $\mathfrak{B}^{\perp}$ and $\mathfrak{B}'^{\perp}$ are equivalent. Choose v in $\mathfrak{B}^{\perp}$ with $n(v) \neq 0$; correspondingly we have v' in $\mathfrak{B}'^{\perp}$ such that $n'(v') = n(v)$. Then $a + vb \to aH_0 + v'(bH_0)$ is an isomorphism of $\mathfrak{B} \perp v\mathfrak{B}$ onto $\mathfrak{B}' \perp v'\mathfrak{B}'$ by the construction above. Hence if we begin with $\mathfrak{B} = F1$, $\mathfrak{B}' = F1'$, repetition of the process gives successively isomorphisms between $\mathfrak{J}$ and $\mathfrak{J}'$, $\mathfrak{Q}$ and $\mathfrak{Q}'$, $\mathfrak{C}$ and $\mathfrak{C}'$.

Remark. We have shown incidentally that, if $\mathfrak{Q}$ is any quaternion sub-algebra containing 1 in a Cayley algebra $\mathfrak{C}$ of characteristic $\neq 2$, then $\mathfrak{Q}$ may be used in the construction of $\mathfrak{C}$ as $\mathfrak{C} = \mathfrak{Q} + v\mathfrak{Q}$. This remark reduces our previous proof of the simplicity of $\mathfrak{C}$ to the following: there is $x \neq 0$ in any nonzero ideal $\mathfrak{B}$ of $\mathfrak{C}$, and x is contained in some quaternion subalgebra $\mathfrak{Q}$ of $\mathfrak{C}$. Then $1 \in \mathfrak{Q} = \mathfrak{Q}x\mathfrak{Q} \subseteq \mathfrak{B}$.

We have already proved, as part of Lemma 3.16, the following Corollary without the restriction on the characteristic of F. We include it here in order to illustrate an elegant method of proof.

Corollary 3.24. *Any two Cayley algebras $\mathfrak{C}$ and $\mathfrak{C}'$ with divisors of zero over F of characteristic $\neq 2$ are isomorphic.*

Proof. Show first that $\mathfrak{C}$ has divisors of zero if and only if there is $w \notin F1$ such that $w^2 = 1$. For $1 - w \neq 0$, $1 + w \neq 0$ imply

$$(1 - w)(1 + w) = 1 - w^2 = 0.$$

Conversely, if $\mathfrak{C}$ has divisors of zero, there exists $x \neq 0$ in $\mathfrak{C}$ with $n(x) = 0$. Then $x = \alpha 1 + u$, u in $(F1)^\perp = \{y \mid t(y) = 0\}$ implies

$$0 = n(x)1 = x\bar{x} = (\alpha 1 + u)(\alpha 1 - u) = \alpha^2 1 - u^2.$$

If $\alpha \neq 0$, then $w = \alpha^{-1}u$ satisfies $w^2 = 1$ ($w \notin F1$). If $\alpha = 0$, then $n(u) = 0$ so that u is an isotropic vector in the nonisotropic space $(F1)^\perp$. Hence $n(x)$ is a universal form and there exists w in $(F1)^\perp$ with $n(w) = -1$ (Jacobson [24], p. 154, Ex. 3), or $w^2 = t(w)w - n(w)1 = 1$ ($w \notin F1$).

Now let $e_1 = \frac{1}{2}(1 - w)$, $e_2 = 1 - e_1 = \frac{1}{2}(1 + w)$. Then $e_1{}^2 = e_1$, $e_2{}^2 = e_2$, $e_1 e_2 = e_2 e_1 = 0$. Also $n(e_i) = 0$ for $i = 1, 2$. Hence every vector in $e_i \mathfrak{C}$ is isotropic since $n(e_i x) = n(e_i)n(x) = 0$. This means that $e_i \mathfrak{C}$ is a totally isotropic subspace ($e_i \mathfrak{C} \subseteq (e_i \mathfrak{C})^\perp$). Hence $\dim (e_i \mathfrak{C}) \leq \frac{1}{2} \dim \mathfrak{C}$ $= 4$ (Jacobson [24], p. 170; Artin [2], p. 122). But $x = 1x = e_1 x + e_2 x$ for all x in $\mathfrak{C}$, so $\mathfrak{C} = e_1 \mathfrak{C} + e_2 \mathfrak{C}$. Hence $\dim (e_i \mathfrak{C}) = 4$, and $n(x)$ has maximal Witt index $= 4 = \frac{1}{2} \dim \mathfrak{C}$. Similarly $n'(x')$ has maximal Witt index $= 4$. Hence $n(x)$ and $n'(x')$ are equivalent (Artin [2], p. 122). By Theorem 3.23, $\mathfrak{C}$ and $\mathfrak{C}'$ are isomorphic.

The following celebrated theorem on quadratic forms permitting composition has been developed through the work of many authors, including Hurwitz [1], Dickson [1], Albert [2], Kaplansky [1], and Jacobson [14]. Generalization to forms of degree n is given in Schafer [19] and McCrimmon [2, 4].

Theorem 3.25 (Hurwitz). *Let $\mathfrak{A}$ be an algebra with 1 (of possibly infinite dimension) over a field F of characteristic $\neq 2$. A necessary and sufficient condition for the existence of a nondegenerate quadratic form $N(x)$ permitting composition on $\mathfrak{A}$ is that $\mathfrak{A}$ be one of the following: $F1$; $F \oplus F$; a separable quadratic field $\mathfrak{Z}$ over F; a quaternion algebra $\mathfrak{Q}$ over F; a Cayley algebra $\mathfrak{C}$ over F. Hence the possible dimensions for $\mathfrak{A}$ are 1, 2, 4, 8. Furthermore, if $\mathfrak{A} = F1$, then $N(\alpha 1) = \alpha^2$; otherwise $N(x)$ is the norm form $n(x)$ given by (3.35).*

Proof. The algebras listed above are called *composition algebras*. We have already proved that any composition algebra $\mathfrak{A}$ has a norm form $n(x)$ satisfying the conditions. For the converse, we assume the existence of a nondegenerate quadratic form $N(x)$ on $\mathfrak{A}$ satisfying

$$(3.63) \qquad N(xy) = N(x)N(y) \qquad \text{for all} \quad x, y \text{ in } \mathfrak{A},$$

and we consider the associated nondegenerate symmetric bilinear form

$$(x, y) = \tfrac{1}{2}[N(x + y) - N(x) - N(y)] \qquad \text{for all} \quad x, y \text{ in } \mathfrak{A}.$$

Then

$$N(x) = (x, x) \qquad \text{for all} \quad x \text{ in } \mathfrak{A},$$

and

$$(xy, xy) = (x, x)(y, y) \qquad \text{for all} \quad x, y \text{ in } \mathfrak{A}.$$

Linearize this relative to x to obtain

$$(xy, zy) = (x, z)N(y) \qquad \text{for all} \quad x, y, z \text{ in } \mathfrak{A}.$$

Linearizing this relative to y, we have

$$(3.64) \quad (xy, zw) + (xw, zy) = 2(x, z)(y, w) \qquad \text{for all} \quad x, y, z, w \text{ in } \mathfrak{A}.$$

Since F has characteristic $\neq 2$, the multilinear identity (3.64) is equivalent to the basic assumption (3.63). Also (3.64) implies

$$(xy, xz) = N(x)(y, z) \qquad \text{for all} \quad x, y, z \text{ in } \mathfrak{A}.$$

Define a trace $T(x)$ on $\mathfrak{A}$ by

$$T(x) = 2(x, 1) \qquad \text{for all} \quad x \text{ in } \mathfrak{A}.$$

Note that $N(1) = 1$ by (3.63) since $N(x) \neq 0$ for some x in $\mathfrak{A}$. Then

$$N(\alpha 1) = \alpha^2, \quad T(\alpha 1) = 2\alpha \qquad \text{for all} \quad \alpha \text{ in } F.$$

Now (3.64) implies

(3.65) $(xy, z) + (x, zy) = (x,z)T(y) \qquad$ for all $\quad x, y, z$ in $\mathfrak{A}$

and

(3.66) $(xy, w) + (y, xw) = T(x)(y, w) \qquad$ for all $\quad x, y, w$ in $\mathfrak{A}$.

Now $(x^2, z) + (x, xz) = T(x)(x, z)$ by (3.66) and $(x, xz) = N(x)(1, z)$. Hence $(x^2 - T(x)x + N(x)1, z) = 0$ for all z in $\mathfrak{A}$. Since (x, y) is non-degenerate, we have

$$x^2 - T(x)x + N(x)1 = 0 \qquad \text{for all} \quad x \text{ in } \mathfrak{A};$$

that is, $\mathfrak{A}$ is a quadratic algebra over F unless $\mathfrak{A} = F1$. By the uniqueness of $n(x)$ in (3.35), we have the final conclusion of the theorem.

Also (3.66) implies that

$$(x(xy), z) = T(x)(xy, z) - (xy, xz) = ((T(x)x - N(x)1)y, z) = (x^2 y, z)$$

for all x, y, z in $\mathfrak{A}$. That is, $x^2 y = x(xy)$ since (x, y) is nondegenerate, implying that $\mathfrak{A}$ is left alternative. Similarly, (3.65) implies that $\mathfrak{A}$ is right alternative.

If $\mathfrak{A}$ is finite-dimensional, then Theorem 3.7 implies that $\mathfrak{A}$ is semi-simple. For, if x in $\mathfrak{A}$ is nilpotent, then $T(x) = 0$ [$x^h = 0$ implies $N(x^h) = [N(x)]^h = 0$; so $N(x) = 0$, whence $x^2 = T(x)x$, and $0 = x^h = [T(x)]^{h-1}x$]. Hence, if x in $\mathfrak{A}$ is properly nilpotent, we have

$$T(xy) = 0 \qquad \text{for all} \quad y \text{ in } \mathfrak{A}.$$

But

$$T(xy) = 2(xy, 1) = 2(x, 1)T(y) - 2(x, y)$$
$$= T(x)T(y) - 2(x, y) = -2(x, y)$$

by (3.65), since x is nilpotent. Hence $(x, y) = 0$ for all y in $\mathfrak{A}$, implying $x = 0$. Thus the radical of $\mathfrak{A}$ is 0, and $\mathfrak{A}$ is semisimple. What are the finite-dimensional semisimple quadratic alternative algebras $\mathfrak{A}$ over F?

If there is more than one simple summand, then $\mathfrak{A} = F \oplus F$. For simple algebras $\mathfrak{A}$, the center is either a separable quadratic extension of F (since the characteristic is $\neq 2$) or $F1$. In the former case, $\mathfrak{A} = 3$; in the latter, $\mathfrak{A}$ is central simple over F. Since $\mathfrak{A}$ is alternative, we have precisely the composition algebras.

Since $\mathfrak{A}$ is assumed to be possibly infinite-dimensional, let $\mathfrak{B}$ be a finite-dimensional subalgebra containing 1. If $\mathfrak{B}$ is nonisotropic, with respect to (x, y), then (x, y) is nondegenerate on $\mathfrak{B}$, and all of the above applies to $\mathfrak{B}$. Furthermore, $\mathfrak{A} = \mathfrak{B} \perp \mathfrak{B}^\perp$ (we have previously given references for this result for finite-dimensional $\mathfrak{A}$; it follows easily from Chapter I and Theorem 3.7 of Artin [2] that the same result holds for finite-dimensional $\mathfrak{B}$ in possibly infinite-dimensional $\mathfrak{A}$). As long as $\mathfrak{B}$ is a proper subalgebra of $\mathfrak{A}$, we may carry out the iterative construction, beginning with $\mathfrak{B} = F1$, of 2^t-dimensional subalgebras of $\mathfrak{A}$. Since $\mathfrak{A}$ is alternative, the process cannot continue beyond the 8-dimensional stage. This completes the proof.

8. DERIVATIONS; SIMPLE LIE ALGEBRAS OF TYPE G

Cayley algebras are intimately associated with the exceptional simple Lie algebras (Jacobson [25], pp. 142–145), and therefore by standard theorems with the exceptional simple Lie groups. We begin by proving that the derivation algebra $\mathfrak{D}(\mathfrak{C})$ of any Cayley algebra $\mathfrak{C}$ over F of characteristic $\neq 2, 3$ is a 14-dimensional central simple Lie algebra (of type G) over F. For this purpose we employ the 28-dimensional orthogonal Lie algebra $\mathfrak{o}(8, n)$ of linear operators (on an 8-dimensional space over F) which are skew relative to the norm form $n(x)$ of $\mathfrak{C}$, and the special linear Lie algebra $\mathfrak{E}'(V)$ of all linear operators of trace 0 on a 3-dimensional vector space V over F.

Let $\mathfrak{A}$ be an alternative algebra over an arbitrary field F. Equations (3.2) imply that

$$(3.67) \qquad [R_x, R_z] = R_{[x,z]} - 2[L_x, R_z] \qquad \text{for all} \quad x, z \text{ in } \mathfrak{A},$$

and

$$(3.68) \qquad [L_x, L_z] = -L_{[x,z]} - 2[L_x, R_z] \qquad \text{for all} \quad x, z \text{ in } \mathfrak{A}.$$

It follows from (3.68) that

$$[L_x, R_z - L_z] = L_{[x,z]} + 3[L_x, R_z] \qquad \text{for all} \quad x, z \text{ in } \mathfrak{A}.$$

We recall that a linear operator D on $\mathfrak{A}$ is a derivation of $\mathfrak{A}$ in case

$$(xy)D = (xD)y + x(yD) \qquad \text{for all} \quad x, y \text{ in } \mathfrak{A};$$

equivalently,

$$[R_y, D] = R_{yD} \qquad \text{for all} \quad y \text{ in } \mathfrak{A},$$

or

$$[L_x, D] = L_{xD} \qquad \text{for all} \quad x \text{ in } \mathfrak{A}.$$

Hence $R_z - L_z$ is a derivation of $\mathfrak{A}$ if and only if $3[L_x, R_z] = 0$ for all x, z in $\mathfrak{A}$. Since $y[L_x, R_z] = (x, y, z)$, we have

Proposition 3.26. *Let $\mathfrak{G}$ be the nucleus of an alternative algebra $\mathfrak{A}$ over F. Then $R_g - L_g$ is a derivation of $\mathfrak{A}$ for all g in $\mathfrak{G}$. Conversely, if the characteristic of F is $\neq 3$, and if $R_g - L_g$ is a derivation of $\mathfrak{A}$, then g is in $\mathfrak{G}$.*

Equations (3.2) also imply that

$$(3.69) \qquad R_x R_y + R_y R_x = R_{xy+yx}, \quad L_x L_y + L_y L_x = L_{xy+yx}$$

for all x, y in $\mathfrak{A}$. [That is, $R(\mathfrak{A})$ and $L(\mathfrak{A})$ are special Jordan algebras.] In the associative algebra $\mathfrak{E} = \mathfrak{E}(\mathfrak{A})$ we have the identity

$$[A, [B, C]] = C(AB + BA) + (AB + BA)C$$

$$- B(AC + CA) - (AC + CA)B$$

for all A, B, C in $\mathfrak{E}$. It follows that

$$[R_y, [R_x, R_z]] = R_z R_{xy+yx} + R_{xy+yx} R_z - R_x R_{yz+zy} - R_{yz+zy} R_x,$$

$$= R_{z(xy)+z(yx)+(xy)z+(yx)z-x(yz)-x(zy)-(yz)x-(zy)x}$$

or

$$[R_y, [R_x, R_z]] = R_{[y,[x,z]]-2(x,y,z)} \qquad \text{for all} \quad x, y, z \text{ in } \mathfrak{A}$$

since $(x, y, z) + (y, x, z) - (y, z, x) - (z, y, x) + (x, z, y) - (z, x, y) = -2(x, y, z)$ in $\mathfrak{A}$. For any x, z in $\mathfrak{A}$, define

(3.70) $$D_{x,z} = R_{[x,z]} - L_{[x,z]} - 3[L_x, R_z].$$

Adding (3.67) and (3.68), we have

$$D_{x,z} = [L_x, L_z] + [L_x, R_z] + [R_x, R_z].$$

If the characteristic of F is $\neq 2$, it follows that $D_{x,z}$ is a derivation of $\mathfrak{A}$ for all x, z in $\mathfrak{A}$. For (3.67) implies that

$$
\begin{aligned}
2[R_y, D_{x,z}] &= 2[R_y, R_{[x,z]} - L_{[x,z]} - 3[L_x, R_z]] \\
&= 3[R_y, R_{[x,z]} - 2[L_x, R_z]] - [R_y, R_{[x,z]}] - 2[R_y, L_{[x,z]}] \\
&= 3[R_y, [R_x, R_z]] - R_{[y,[x,z]]} \\
&= R_{2[y,[x,z]] - 6(x,y,z)} \\
&= 2R_{yD_{x,z}}
\end{aligned}
$$

for all x, y, z in $\mathfrak{A}$ since $[L_a, R_b] = [R_a, L_b]$ by (3.2). Any sum

(3.71) $$D = \sum D_{x_i, z_i}, \qquad x_i, z_i \text{ in } \mathfrak{A},$$

is a derivation of $\mathfrak{A}$, if the characteristic of F is $\neq 2$. The derivations in Proposition 3.26 and in (3.71) are inner by the definition in Chapter II. We have seen that the inner derivations form an ideal of $\mathfrak{D}(\mathfrak{A})$. Here we shall have occasion to use the explicit formula

(3.72) $$[D_{x,z}, D] = D_{xD,z} + D_{x,zD}$$

for all D in $\mathfrak{D}(\mathfrak{A})$; x, z in $\mathfrak{A}$; $\mathfrak{D}_{x,z}$ in (3.70).

The Lie multiplication algebra $\mathfrak{L}(\mathfrak{A})$ of any alternative algebra $\mathfrak{A}$ over F of characteristic $\neq 2$ is

$$\mathfrak{L}(\mathfrak{A}) = R(\mathfrak{A}) + L(\mathfrak{A}) + [L(\mathfrak{A}), R(\mathfrak{A})].$$

For, just as we could conclude above from the first part of (3.69) that $[R(\mathfrak{A}), [R(\mathfrak{A}), R(\mathfrak{A})]] \subseteq R(\mathfrak{A})$, we have also $[L(\mathfrak{A}), [L(\mathfrak{A}), L(\mathfrak{A})]] \subseteq L(\mathfrak{A})$. Then (3.67) and (3.68) imply that

$$[R(\mathfrak{A}), [L(\mathfrak{A}), R(\mathfrak{A})]] \subseteq [R(\mathfrak{A}), R(\mathfrak{A})] + R(\mathfrak{A})$$

and

$$[L(\mathfrak{A}), [L(\mathfrak{A}), R(\mathfrak{A})]] \subseteq [L(\mathfrak{A}), L(\mathfrak{A})] + L(\mathfrak{A}).$$

In Chapter II we saw that

$$\mathfrak{L}(\mathfrak{A}) = \sum \mathfrak{H}_i \qquad \text{with} \qquad \mathfrak{H}_1 = R(\mathfrak{A}) + L(\mathfrak{A}), \qquad \mathfrak{H}_i = [\mathfrak{H}_1, \mathfrak{H}_{i-1}].$$

But here $\mathfrak{H}_i \subseteq \mathfrak{H}_1 + \mathfrak{H}_2$ for $i \geq 3$, so that

$$\mathfrak{L}(\mathfrak{A}) = \mathfrak{H}_1 + \mathfrak{H}_2 = R(\mathfrak{A}) + L(\mathfrak{A}) + [L(\mathfrak{A}), R(\mathfrak{A})].$$

It follows that, if $\mathfrak{A}$ contains 1, and if F is also of characteristic $\neq 3$, then any inner derivation of $\mathfrak{A}$ is the sum of a derivation (3.71) and a derivation of the type in Proposition 3.26. By (3.70) any element of $\mathfrak{L}(\mathfrak{A})$ has the form

$$D = R_g + L_h + \sum D_{x_i, z_i}.$$

Then, if D is a derivation,

$$0 = 1D = g + h + \sum [1, [x_i, z_i]] - 3 \sum (x_i, 1, z_i) = g + h,$$

so that

$$D - \sum D_{x_i, z_i} = R_g - L_g$$

is a derivation of $\mathfrak{A}$. By Proposition 3.26 we have g in the nucleus $\mathfrak{G}$ of $\mathfrak{A}$, and

$$D = R_g - L_g + \sum_i D_{x_i, z_i}, \qquad g \text{ in } \mathfrak{G}, \quad x_i, z_i \text{ in } \mathfrak{A},$$

as desired.

We next prove that

(3.73) $D_{xy,z} + D_{yz,x} + D_{zx,y} = 0 \qquad \text{for all} \quad x, y, z \text{ in } \mathfrak{A}.$

Identity (2.4) implies

$$a(x, y, z) + (a, x, y)z = \quad (ax, y, z) - (a, xy, z) + (a, x, yz),$$

$$-x(y, z, a) - (x, y, z)a = -(xy, z, a) + (x, yz, a) - (x, y, za),$$

$$x(a, y, z) + (x, a, y)z = \quad (xa, y, z) - (x, ay, z) + (x, a, yz).$$

Adding these, we have

$$[a, (x, y, z)] = (ax + xa, y, z) - 2(a, xy, z)$$
$$- (a, yz, x) - (za, x, y) - (ay, z, x)$$

in any alternative algebra $\mathfrak{A}$. Let $\sum$ denote the sum of three terms obtained by cyclic permutation of x, y, z. Then

$$3[a, (x, y, z)] = \sum [a, (x, y, z)]$$
$$= \sum (ax + xa, y, z) - 2\sum (a, xy, z) - \sum (a, xy, z)$$
$$- \sum (xa, y, z) - \sum (ax, y, z)$$
$$= 3\sum (xy, a, z),$$

so that

$$3(R_{(x,y,z)} - L_{(x,y,z)} - [L_{xy}, R_z] - [L_{yz}, R_x] - [L_{zx}, R_y]) = 0$$

for all x, y, z in $\mathfrak{A}$. Now

(3.74) $[xy, z] + [yz, x] + [zx, y] = 3(x, y, z)$ for all x, y, z in $\mathfrak{A}$.

Hence

$$D_{xy,z} + D_{yz,x} + D_{zx,y} = 3R_{(x,y,z)} - 3L_{(x,y,z)} - 3[L_{xy}, R_z]$$
$$- 3[L_{yz}, R_x] - 3[L_{zx}, R_y] = 0,$$

implying (3.73).

Let $\mathfrak{A}$ be a composition algebra over F of characteristic $\neq 2$, $t(x)$ and $n(x)$ the trace and norm forms on $\mathfrak{A}$, and (x, y) in (3.60) be the associated nondegenerate symmetric bilinear form. We have seen in (3.62) that $t(xy)$ is a trace form on $\mathfrak{A}$. Now $\mathfrak{A} = F1 \perp \mathfrak{A}_0$ where $\mathfrak{A}_0 = (F1)^{\perp}$ is the space of all elements of trace 0. If $\dim \mathfrak{A} \geq 4$, then $\mathfrak{A}_0$ may also be characterized as

$$\mathfrak{A}_0 = \{\sum [x, y] \mid x, y \text{ in } \mathfrak{A}\}.$$

For $z = \sum [x, y]$ implies $t(z) = \sum t([x, y]) = 0$ while, conversely, any element of trace 0 in $\mathfrak{A} = \mathfrak{B} + v\mathfrak{B}$ is a sum of commutators, since b in $\mathfrak{B}$ implies

$$b - \bar{b} = [\mu^{-1}v, vb], \qquad v = [vv_1, \tfrac{1}{2}\mu_1^{-1}v_1], \qquad v(b - \bar{b}) = [v, b]$$

by (3.38). Let D be any derivation of $\mathfrak{A}$; then

(3.75) $\mathfrak{A}D \subseteq \mathfrak{A}_0.$

For $\mathfrak{A}D = 0$ if dim ≤ 2, while in case dim $\mathfrak{A} \geq 4$ we have $1D = 0$ and $z = \sum [x, y]$ in $\mathfrak{A}_0$, implying $zD = \sum [xD, y] + \sum [x, yD]$ is in $\mathfrak{A}_0$. It follows that

$$\overline{xD} = -xD = \bar{x}D \qquad \text{for all} \quad x \text{ in } \mathfrak{A},$$

since xD in $\mathfrak{A}_0$ implies $xD + \overline{xD} = 0$, while $(x + \bar{x})D = t(x)1D = 0$. It follows also that

$$(xD, y) + (x, yD) = 0 \qquad \text{for all} \quad x, y \text{ in } \mathfrak{A}.$$

For

$$2(xD, y) + 2(x, yD) = t((xD)\bar{y} + x(\overline{yD})) = t((xD)\bar{y} + x(\bar{y}D))$$
$$= t((x\bar{y})D) = 0.$$

Since (x, y) is nondegenerate on $\mathfrak{A}$, for each T in $\mathfrak{E}(\mathfrak{A})$ there is a unique *adjoint* T^* in $\mathfrak{E}(\mathfrak{A})$ satisfying

$$(xT, y) = (x, yT^*) \qquad \text{for all} \quad x, y \text{ in } \mathfrak{A};$$

T is *skew*, relative to $n(x)$ and to (x, y), in case

$$T^* = -T.$$

Hence

$$D^* = -D \qquad \text{for all} \quad D \text{ in } \mathfrak{D}(\mathfrak{A}).$$

Since $n(x)$ permits composition, we have already established (3.65) with $T(y)$ replaced by $t(y)$: $(xR_y, z) = (x, zR_{\bar{y}})$ for all x, y, z in $\mathfrak{A}$; that is,

$$R_y^* = R_{\bar{y}} \qquad \text{for all} \quad y \text{ in } \mathfrak{A}.$$

Similarly, (3.66) is equivalent to

$$L_x^* = L_{\bar{x}} \qquad \text{for all} \quad x \text{ in } \mathfrak{A}.$$

For elements of trace 0, we have

$$R_s^* = -R_s \qquad \text{for all} \quad s \text{ in } \mathfrak{A}_0$$

and

$$L_t^* = -L_t \qquad \text{for all} \quad t \text{ in } \mathfrak{A}_0.$$

Let $\mathfrak{C}$ be a Cayley algebra over F. Since $\mathfrak{C}$ is 8-dimensional, it is well known that the set $\mathfrak{o}(8, n)$ of all skew elements of $\mathfrak{E} = \mathfrak{E}(\mathfrak{C})$ is an orthogonal Lie algebra of dimension $\frac{1}{2} \cdot 8 \cdot (8 - 1) = 28$ over F. Also we have seen that $\mathfrak{o}(8, n)$ contains all

$$D + R_s + L_t, \qquad D \text{ in } \mathfrak{D}(\mathfrak{C}); \quad s, t \text{ in } \mathfrak{C}_0.$$

Suppose that $D + R_s + L_t = 0$. Applying this to 1 in $\mathfrak{C}$, we have $0 = 1D + s + t$, or $t = -s$. Then $R_s - L_s = -D$ is a derivation of $\mathfrak{C}$. Hence, if we make the additional assumption that the characteristic of F is $\neq 3$, Proposition 3.26 implies that s is in the nucleus $F1$ of $\mathfrak{C}$. That is, s is in $F1 \cap \mathfrak{A}_0$, so $s = 0$, implying $t = 0$ and $D = 0$. If we denote by $R_0(\mathfrak{C})$ (resp. $L_0(\mathfrak{C})$) the set of all right (resp. left) multiplications of $\mathfrak{C}$ corresponding to elements of trace 0, we have a vector space direct sum $\mathfrak{D}(\mathfrak{C}) + R_0(\mathfrak{C}) + L_0(\mathfrak{C}) \subseteq \mathfrak{o}(8, n)$. Since $\mathfrak{C}_0$ is 7-dimensional, $R_0(\mathfrak{C})$ and $L_0(\mathfrak{C})$ are also. Then dim $\mathfrak{o}(8, n) = 28$ implies dim $\mathfrak{D}(\mathfrak{C}) \leq 14$. We shall have proved

$$(3.76) \qquad \mathfrak{o}(8, n) = \mathfrak{D}(\mathfrak{C}) + R_0(\mathfrak{C}) + L_0(\mathfrak{C}) \qquad \text{(direct sum)}$$

when we establish dim $\mathfrak{D}(\mathfrak{C}) = 14$.

In the familiar classification of finite-dimensional simple Lie algebras over an algebraically closed field of characteristic 0 (Jacobson [25], pp. 135–146), there is exactly one 14-dimensional algebra, the exceptional algebra G_2. We say that a (central simple) Lie algebra $\mathfrak{L}$ over F of characteristic 0 is of *type G* in case, for the algebraic closure K of F, we have $\mathfrak{L}_K \cong G_2$. In Theorem 3.28 we shall see that $G_2 \cong \mathfrak{D}(\mathfrak{C})$ for $\mathfrak{C}$ the (split) Cayley algebra over K. For F of characteristic $\neq 2$, 3 we extend the definition above, defining $\mathfrak{L}$ to be of *type G* in case, for the algebraic closure K of F, we have $\mathfrak{L}_K \cong \mathfrak{D}(\mathfrak{C})$ where $\mathfrak{C}$ is the (split) Cayley algebra over K.

Lemma 3.27. *Let* $\mathfrak{A}$ *be a finite-dimensional nonassociative algebra over an arbitrary field* F *satisfying*

(i) $\mathfrak{A}$ *is a vector space direct sum* $\mathfrak{A} = \mathfrak{S} + \mathfrak{T}$ *where* dim $\mathfrak{S} >$ dim $\mathfrak{T}$;

(ii) $\mathfrak{S}$ *is a simple subalgebra of* $\mathfrak{A}$;

(iii) $\mathfrak{S}\mathfrak{T} = \mathfrak{T} \supseteq \mathfrak{T}\mathfrak{S}$ (*or* $\mathfrak{T}\mathfrak{S} = \mathfrak{T} \supseteq \mathfrak{S}\mathfrak{T}$).

If $\mathfrak{B}$ *is a proper ideal of* $\mathfrak{A}$, *then* $\mathfrak{B} \subseteq \mathfrak{T}$.

Proof. Since $\mathfrak{B} \cap \mathfrak{S}$ is an ideal of $\mathfrak{S}$, we have either $\mathfrak{B} \cap \mathfrak{S} = \mathfrak{S}$ or $\mathfrak{B} \cap \mathfrak{S} = 0$ by (ii). In the first case $\mathfrak{B} \supseteq \mathfrak{S}$. But then $\mathfrak{B} \supseteq \mathfrak{B} \mathfrak{T} \supseteq \mathfrak{S} \mathfrak{T} = \mathfrak{T}$ by (iii), implying $\mathfrak{B} \supseteq \mathfrak{S} + \mathfrak{T} = \mathfrak{A}$, a contradiction. Hence $\mathfrak{B} \cap \mathfrak{S} = 0$. Let $\mathfrak{S}_{\mathfrak{B}}$ be the image of $\mathfrak{B}$ under the projection of $\mathfrak{A}$ onto $\mathfrak{S}$ relative to the direct sum decomposition $\mathfrak{A} = \mathfrak{S} + \mathfrak{T}$, that is,

$$\mathfrak{S}_{\mathfrak{B}} = \{ s \in \mathfrak{S} \mid \exists\, t \in \mathfrak{T}, \; s + t \in \mathfrak{B} \}.$$

Then $\mathfrak{S}_{\mathfrak{B}}$ is a right ideal of $\mathfrak{S}$. For s in $\mathfrak{S}_{\mathfrak{B}}, s'$ in $\mathfrak{S}$, imply there exists t in $\mathfrak{T}$ such that $s + t$ is in $\mathfrak{B}$, so $\mathfrak{B}$ contains $(s + t)s' = ss' + ts'$ where ss' is in $\mathfrak{S}$, ts' in $\mathfrak{T}$ by (iii); hence ss' is in $\mathfrak{S}_{\mathfrak{B}}$ as desired. Similarly $\mathfrak{S}_{\mathfrak{B}}$ is a left ideal of $\mathfrak{B}$. Hence $\mathfrak{S}_{\mathfrak{B}} = \mathfrak{S}$ or $\mathfrak{S}_{\mathfrak{B}} = 0$. The first case leads to a contradiction. For $\mathfrak{S}_{\mathfrak{B}} = \mathfrak{S}$ implies that, for every s in $\mathfrak{S}$, there exists t in $\mathfrak{T}$ such that $s + t$ is in $\mathfrak{B}$. Let $s_1, \ldots, s_n$ be a basis for $\mathfrak{S}$ over F. Then there exist $t_1, \ldots, t_n$ in $\mathfrak{T}$, satisfying $s_i + t_i \in \mathfrak{B}$. Hence there is a linear mapping U of $\mathfrak{S}$ into $\mathfrak{T}$ (extending $s_i U = t_i$) which satisfies

$$s + sU \quad \text{is in} \quad \mathfrak{B} \qquad \text{for all} \quad s \text{ in } \mathfrak{S}.$$

Since $\dim \mathfrak{S} > \dim \mathfrak{T}$ by (i), there exists $s \neq 0$ in $\mathfrak{S}$ such that $sU = 0$. For this s we have $0 \neq s \in \mathfrak{B} \cap \mathfrak{S} = 0$, a contradiction. Hence $\mathfrak{S}_{\mathfrak{B}} = 0$. That is, $\mathfrak{B} \subseteq \mathfrak{T}$.

Theorem 3.28   (Cartan–Jacobson).   *Let $\mathfrak{C}$ be a Cayley algebra over F of characteristic $\neq 2, 3$. Then the derivation algebra $\mathfrak{D}(\mathfrak{C})$ of $\mathfrak{C}$ is a 14-dimensional central simple Lie algebra (of type G).*

Proof.   Since $\mathfrak{D}(\mathfrak{C})_K = \mathfrak{D}(\mathfrak{C}_K)$ for any extension K of F, it is sufficient to prove that $\dim \mathfrak{D}(\mathfrak{C}) = 14$ and that $\mathfrak{D}(\mathfrak{C})$ is simple in case $\mathfrak{C}$ is the split Cayley algebra. We take $\mathfrak{C}$ in the form given by Lemma 3.16. Denote by V (resp. W) the 3-dimensional space $\mathfrak{C}_{12}$ (resp. $\mathfrak{C}_{21}$) $\subseteq \mathfrak{C}_0$. Since $t(e_1) = t(e_2) = 1$, it follows from (3.60) and the proof of Lemma 3.16 that

$$(3.77) \quad uw = \langle u, w \rangle e_1, \qquad wu = \langle u, w \rangle e_2 \qquad \text{for all} \quad u \text{ in } V, w \text{ in } W,$$

where

$$(3.78) \qquad \langle u, w \rangle = t(uw) = -2(u, w) \qquad (u \text{ in } V, \quad w \text{ in } W)$$

is a nondegenerate bilinear form connecting V and W. Since b, c in V implies bc is in W by (3.18), we may consider the trilinear form

(3.79) $\langle a, bc \rangle$, a, b, c in V; $\langle u, w \rangle$ in (3.78).

Now (3.79) is an alternating trilinear form on the 3-dimensional space V: $\langle b, a^2 \rangle = \langle a, ab \rangle = \langle a, ba \rangle = 0$ for all a, b in V. For $a^2 = 0$ by (3.20), and $\langle a, ab \rangle = t(a(ab)) = t(a^2 b) = 0$, while $\langle a, ba \rangle = -\langle a, ab \rangle = 0$ by (3.21). It follows that $\langle a, bc \rangle$ is a scalar multiple of the determinant (Artin [1], pp. 11–15):

$$\langle a, bc \rangle = \varepsilon \det(a; b; c), \qquad a, b, c \text{ column vectors in } V.$$

Also $\varepsilon \neq 0$ since $\langle u, w \rangle$ is nondegenerate: $\langle a, bc \rangle = 0$ for all a, b, c in V implies $bc = 0$ for all b, c in V; then e_{12}, ve_2 in $\mathfrak{C}_{12} = V$ implies $0 = e_{12}(ve_2) = v(\bar{e}_{12}e_2) = -ve_{12} \neq 0$, a contradiction. It is an elementary property of determinants of order n that, if F contains more than n elements and if T is any linear operator on the n-dimensional space V of column vectors, then

$$\sum_{i=1}^{n} \det(a_1; \ldots; a_i T; \ldots; a_n) = \det(a_1; \ldots; a_n) \text{ trace } T, \qquad a_i \text{ in } V.$$

In particular, for any T of trace 0 the left-hand side is 0, so that $\det(a_1; \ldots; a_n)$ is called (*Lie*) *invariant* under any T in $\mathfrak{C}'(V)$, the $(n^2 - 1)$-dimensional subspace (Lie subalgebra) of all elements of trace 0 in $\mathfrak{C}(V)$. For, if $\tilde{T}$ is the matrix corresponding to T, then

$$A = (a_1; a_2; \ldots; a_n) \qquad \text{implies} \qquad A\tilde{T} = (a_1 T; a_2 T; \ldots; a_n T),$$

and we may equate the coefficients of λ in two expressions for $\det(A + \lambda A\tilde{T})$ as follows:

$$\det(A + \lambda A\tilde{T}) = \det(a_1 + \lambda a_1 T; a_2 + \lambda a_2 T; \ldots; a_n + \lambda a_n T)$$
$$\equiv \det A + \lambda \left[\sum_{i=1}^{n} \det(a_1; \ldots; a_i T; \ldots; a_n) \right] (\text{mod } \lambda^2)$$

and

$$\det(A + \lambda A\tilde{T}) = (\det A)\det(I + \lambda \tilde{T}) = \det A(1 + \lambda \text{ trace } \tilde{T} + \cdots)$$
$$\equiv \det A + \lambda(\det A)(\text{trace } T) (\text{mod } \lambda^2).$$

In our situation there are more than three elements in F, so

$$\det(aT; b; c) + \det(a; bT; c) + \det(a; b; cT) = \det(a; b; c) \text{ trace } T$$

for all a, b, c in our 3-dimensional V, T in $\mathfrak{E}(V)$. Using $\varepsilon \neq 0$, we have

$$\langle aT, bc \rangle + \langle a, (bT)c \rangle + \langle a, b(cT) \rangle = 0$$

for all a, b, c in V, T in $\mathfrak{E}'(V)$. If T^* denotes the adjoint of T with respect to the nondegenerate form $\langle u, w \rangle$ connecting V and W, we have

$$\langle a, (bc)T^* \rangle + \langle a, (bT)c \rangle + \langle a, b(cT) \rangle = 0,$$

or

(3.80) $$(bc)T^* + (bT)c + b(cT) = 0$$

for all b, c in V, T in $\mathfrak{E}'(V)$. By (3.77) we may interchange e_1 and e_2 to obtain the dual relationship

(3.81) $$(bc)T + (bT^*)c + b(cT^*) = 0$$

for all b, c in W, T in $\mathfrak{E}'(V)$, since $T \in \mathfrak{E}'(V)$ is equivalent to $T^* \in \mathfrak{E}'(W)$.
  Let

$$\mathfrak{D}_0 = \{ D \mid D \in \mathfrak{D}(\mathfrak{C}), \; e_1 D = 0 \}.$$

For any T in $\mathfrak{E}'(V)$, define D_T by

$$(\alpha e_1 + u + w + \beta e_2)D_T = uT - wT^*;$$

α, β in F, u in V, w in W. Then D_T is in $\mathfrak{D}_0$. For, if $x = \alpha e_1 + u + w + \beta e_2$ and $y = \gamma e_1 + u' + w' + \delta e_2$ (u' in V, w' in W), then

$$xy = (\alpha\gamma + \langle u, w' \rangle)e_1 + (\alpha u' + \delta u + ww')$$
$$+ (\gamma w + \beta w' + uu') + (\beta\delta + \langle u', w \rangle)e_2.$$

Hence

$$(xD_T)y + x(yD_T) = (\langle uT, w' \rangle - \langle u, w'T^* \rangle)e_1$$
$$+ (\delta uT - (wT^*)w' + \alpha u'T - w(w'T^*))$$
$$+ ((uT)u' - \gamma wT^* + u(u'T) - \beta w'T^*)$$
$$+ (-\langle u', wT^* \rangle + \langle u'T, w \rangle)e_2$$
$$= (\alpha u' + \delta u + ww')T - (\gamma w + \beta w' + uu')T^*$$
$$= (xy)D_T$$

by (3.80) and (3.81), so that D_T is in $\mathfrak{D}(\mathbb{C})$ for all T in $\mathfrak{E}'(V)$. Also $e_1 D_T = 0$, implying D_T is in $\mathfrak{D}_0$. Since $D_T = 0$ implies $T = 0$, we have

$$\dim \mathfrak{D}_0 \geq \dim \mathfrak{E}'(V) = 8.$$

Now $\mathfrak{D}(\mathbb{C})$ contains all derivations of the form

$$\tilde{D} + D_{e_1,u} + D_{e_2,w}, \qquad \tilde{D} \text{ in } \mathfrak{D}_0, \quad u \text{ in } V, \quad w \text{ in } W,$$

where $D_{x,z}$ is defined by (3.70). Suppose that $D = \tilde{D} + D_{e_1,u} + D_{e_2,w} = 0$. Then $0 = e_1 D = e_1 D_{e_1,u} + e_1 D_{e_2,w} = [e_1, [e_1, u]] - 3(e_1, e_1, u) + [e_1, [e_2, w]] - 3(e_2, e_1, w) = u - w$, implying $u = w = 0$. Hence $D_{e_1,u} = D_{e_2,w} = \tilde{D} = 0$, and we have a vector space direct sum $\mathfrak{D}_0 + \mathfrak{D}_1 + \mathfrak{D}_2 \subseteq \mathfrak{D}(\mathbb{C})$ where

$$\mathfrak{D}_1 = \{D_{e_1,u} \mid u \in V\}, \qquad \mathfrak{D}_2 = \{D_{e_2,w} \mid w \in W\}.$$

Since $\dim \mathfrak{D}_i = 3$ ($i = 1, 2$), it follows that $8 + 6 \leq \dim \mathfrak{D}_0 + 6 \leq \dim \mathfrak{D}(\mathbb{C}) \leq 14$, implying $\dim \mathfrak{D}(\mathbb{C}) = 14$ and $\mathfrak{o}(8, n) = \mathfrak{D}(\mathbb{C}) + R_0(\mathbb{C}) + L_0(\mathbb{C})$. Also

(3.82) $$\mathfrak{D}(\mathbb{C}) = \mathfrak{D}_0 + \mathfrak{D}_1 + \mathfrak{D}_2 \qquad \text{(direct sum)}.$$

It is well known (and easy to verify) that, since the characteristic of F is $\neq 3$, the Lie algebra $\mathfrak{E}'(V)$ is simple. Also, since the associative enveloping algebra of $\mathfrak{E}'(V)$ is $\mathfrak{E}(V)$, we have

(3.83) $$V\mathfrak{E}'(V) = V.$$

We use these facts to show that $\mathfrak{D}(\mathbb{C})$ is simple. Now $T \to D_T$ gives an isomorphism between the Lie algebras $\mathfrak{E}'(V)$ and $\mathfrak{D}_0$; hence $\mathfrak{D}_0$ is simple. Since $e_2 D = (1 - e_1)D = 0$ for all D in $\mathfrak{D}_0$, it follows from (3.72) that

(3.84) $$[D_{e_1,u}, D] = D_{e_1,uD}, \qquad [D_{e_2,w}, D] = D_{e_2,wD}$$

for all D in $\mathfrak{D}_0$, u in V, w in W. Also u in V, D in $\mathfrak{D}_0$ imply $uD = (e_1 u e_2)D = e_1(uD)e_2$ so that $V\mathfrak{D}_0 \subseteq V$; similarly, $W\mathfrak{D}_0 \subseteq W$. Hence (3.83) and (3.84), together with the duality between V and W, imply

$$[\mathfrak{D}_0, \mathfrak{D}_1] = \mathfrak{D}_1, \qquad [\mathfrak{D}_0, \mathfrak{D}_2] = \mathfrak{D}_2.$$

We shall also use the relationships

(3.85) $$[\mathfrak{D}_1, \mathfrak{D}_1] \subseteq \mathfrak{D}_2, \qquad [\mathfrak{D}_2, \mathfrak{D}_2] \subseteq \mathfrak{D}_1.$$

For u, u' in V, we have

$$uD_{e_1,u'} = [u, [e_1, u']] - 3(e_1, u, u') = uu' - u'u - 3uu'$$

$$= -(uu' + u'u) - uu' = -uu'$$

by (3.18) and (3.21), and

$$e_1 D_{e_1,u'} = [e_1, [e_1, u']] - 3(e_1, e_1, u') = u'.$$

Hence

$$[D_{e_1,u}, D_{e_1,u'}] = D_{u',u} - D_{e_1,uu'} = D_{u',u} + D_{e_2,uu'}$$

by (3.72). But (3.73) implies that

$$D_{e_2,uu'} = -D_{uu',e_2} = D_{u'e_2,u} + D_{e_2u,u'} = D_{u',u},$$

so that

$$(3.86) \qquad [D_{e_1,u}, D_{e_1,u'}] = 2D_{e_2,uu'} \qquad \text{for all} \quad u, u' \text{ in } V.$$

This establishes the first part of (3.85). Symmetrically, we have the second part of (3.85).

Let $\mathfrak{B}$ be an ideal of $\mathfrak{D}(\mathbb{C})$, $\mathfrak{B} \neq \mathfrak{D}(\mathbb{C})$. We may apply Lemma 3.27 with $\mathfrak{A} = \mathfrak{D}(\mathbb{C})$, $\mathfrak{S} = \mathfrak{D}_0$ of dimension 8, $\mathfrak{T} = \mathfrak{D}_1 + \mathfrak{D}_2$ of dimension 6, $[\mathfrak{D}_0, \mathfrak{D}_1 + \mathfrak{D}_2] = \mathfrak{D}_1 + \mathfrak{D}_2$, to see that $\mathfrak{B}$ is contained in $\mathfrak{D}_1 + \mathfrak{D}_2$. Let

$$D' = D_{e_1,u''} + D_{e_2,w}$$

be any element of $\mathfrak{B}$, and take any $u' \neq 0$ in V. Then

$$[D', D_{e_1,u'}] = [D_{e_1,u''}, D_{e_1,u'}] + [D_{e_2,w}, D_{e_1,u'}]$$

is in $\mathfrak{B} \subseteq \mathfrak{D}_1 + \mathfrak{D}_2$, so that (3.85) implies $[D_{e_2,w}, D_{e_1,u'}]$ is in $\mathfrak{D}_1 + \mathfrak{D}_2$. But

$$e_2 D_{e_1,u'} = [e_2, [e_1, u']] - 3(e_1, e_2, u') = -u'$$

and

$$wD_{e_1,u'} = [w, [e_1, u']] - 3(e_1, w, u')$$

$$= [w, u'] = \langle u', w \rangle (e_2 - e_1) = \langle u', w \rangle (2e_2 - 1)$$

by (3.77). Hence (3.72) implies that

$$[D_{e_2,w}, D_{e_1,u'}] = -D_{u',w} + \langle u', w \rangle D_{e_2,2e_2-1} = -D_{u',w}$$

is in $\mathfrak{D}_0$ since

$$e_1 D_{u',w} = [e_1, [u', w]] - 3(u', e_1, w)$$
$$= \langle u', w \rangle [e_1, e_1 - e_2] = 0.$$

Hence $D_{u',w}$ is in $\mathfrak{D}_0 \cap (\mathfrak{D}_1 + \mathfrak{D}_2) = 0$. For every u in V, we have

$$0 = u D_{u',w} = [u, [u', w]] - 3(u', u, w)$$
$$= \langle u', w \rangle [u, e_1 - e_2] + 3(u', w, u) = \langle u', w \rangle u - 3 \langle u, w \rangle u',$$

or

(3.87) $\langle u', w \rangle u - 3 \langle u, w \rangle u' = 0$ for all u in V.

Let $u = u'$ in (3.87). Then $-2 \langle u', w \rangle u' = 0$, or $\langle u', w \rangle = 0$ since $u' \neq 0$. That is, (3.87) implies $\langle u, w \rangle = 0$ for all u in V. Since $\langle u, w \rangle$ is non-degenerate, we have $w = 0$. That is, any element D' of $\mathfrak{B}$ is in $\mathfrak{D}_1$. Symmetrically, D' is in $\mathfrak{D}_2$. Hence $\mathfrak{B} \subseteq \mathfrak{D}_1 \cap \mathfrak{D}_2 = 0$. We have shown that any ideal of $\mathfrak{D}(\mathfrak{C})$ is either $\mathfrak{D}(\mathfrak{C})$ itself or 0; $\mathfrak{D}(\mathfrak{C})$ is simple.

Corollary 3.29. *Every derivation D of a Cayley algebra $\mathfrak{C}$ of characteristic* $\neq 2, 3$ *is inner:*

$$D = \sum D_{x_i, z_i} \quad \text{for} \quad x_i, z_i \text{ in } \mathfrak{C}.$$

Proof. By (3.72) the derivations $\sum D_{x_i, z_i}$ form an ideal in the simple algebra $\mathfrak{D}(\mathfrak{C})$. This ideal is $\neq 0$, since it is sufficient to prove this over an algebraically closed field. If $\mathfrak{C}$ is a split Cayley algebra, we have seen in the proof of Theorem 3.28 that $e_1 D_{e_1, u} = u \neq 0$ for any $u \neq 0$ in V.

We omit the proof of the fact that every central simple Lie algebra $\mathfrak{L}$ of type G over F (of characteristic $\neq 2, 3$) is isomorphic to $\mathfrak{D}(\mathfrak{C})$ for some Cayley algebra $\mathfrak{C}$ over F, and that $\mathfrak{D}(\mathfrak{C}_1) \cong \mathfrak{D}(\mathfrak{C}_2)$ if and only if $\mathfrak{C}_1 \cong \mathfrak{C}_2$ (Jacobson [3]; Barnes [1]; Seligman [3]).

Corollary 3.30. *Every derivation D of a finite-dimensional separable alternative algebra $\mathfrak{A}$ of characteristic* $\neq 2, 3$ *is inner:*

(3.88) $D = R_g - L_g + \sum D_{x_i, z_i}, \qquad g \text{ in } \mathfrak{G}; \quad x_i, z_i \text{ in } \mathfrak{A};$

where $\mathfrak{G}$ is the nucleus of $\mathfrak{A}$. If $\mathfrak{A}$ is of characteristic 0, we may take $g = 0$ in (3.88).

Proof.   As in the proof of Theorem 2.5, it is sufficient to prove this for (separable) simple algebras $\mathfrak{A}$. Since $\mathfrak{A}$ contains 1, the center K of $\mathfrak{A}$ is a finite separable extension of F, $\mathfrak{A}$ is central simple over K, and the multiplication center is $K^* = \{R_k \,|\, k \text{ in } K\}$ by Theorem 2.1. Since K is a characteristic subalgebra, any derivation D of $\mathfrak{A}$ induces a derivation on K, which is 0 by the elementary fact already quoted preceding Theorem 2.5. Hence $[R_x, D] = R_{xD}$ implies

$$(3.89) \qquad [R_k, D] = 0 \qquad \text{for all} \quad k \text{ in } K; \quad D \text{ in } \mathfrak{D}(\mathfrak{A}).$$

For k in K, D in $\mathfrak{D}(\mathfrak{A})$, define a new kD by $kD = R_k D$. Then (3.89) implies that kD is in $\mathfrak{D}(\mathfrak{A})$, that $\mathfrak{D}(\mathfrak{A})$ is a vector space over K, and that $k[D, D'] = [kD, D'] = [D, kD']$ for all k in K; D, D' in $\mathfrak{D}(\mathfrak{A})$. That is, $\mathfrak{D}(\mathfrak{A})$ is an algebra over K. Equation (3.89) yields $(kx)D = k(xD)$ for all k in K, x in $\mathfrak{A}$, so that any derivation D of $\mathfrak{A}$ over F is not merely an F-endomorphism of $\mathfrak{A}$ but a K-endomorphism of $\mathfrak{A}$. That is, $\mathfrak{D}(\mathfrak{A} \text{ over } K) = \mathfrak{D}(\mathfrak{A})$ over K. If $\mathfrak{A}$ over K is a Cayley algebra, then every derivation of $\mathfrak{A}$ has the form $\sum D_{x_i, z_i}$ by Corollary 3.29. If $\mathfrak{A}$ is a central simple associative algebra over K, it is well known (Jacobson [2]) that for arbitrary characteristic every derivation is of the form $R_g - L_g$. Hence (3.88) holds. It remains to remark that, for any g in a central simple associative algebra $\mathfrak{A}$ over K of characteristic 0, we have α in K and x_i, z_i in $\mathfrak{A}$ satisfying $g = \alpha 1 + \sum [x_i, z_i]$, so that $R_g - L_g = \sum D_{x_i, z_i}$ by (3.70).

In Chapter IV we shall have occasion to use the following theorem.

Theorem 3.31 (Principle of triality). *Let $\mathfrak{C}$ be a Cayley algebra of characteristic $\neq 2$, 3 with norm $n(x)$, and let $\mathfrak{o}(8, n)$ be the orthogonal Lie algebra of all U in $\mathfrak{E}(\mathfrak{C})$ which are skew relative to $n(x)$. For every U in $\mathfrak{o}(8, n)$ there are unique U', U'' in $\mathfrak{o}(8, n)$ satisfying*

$$(3.90) \qquad (xy)U = (xU')y + x(yU'') \qquad \text{for all} \quad x, y \text{ in } \mathfrak{C}.$$

Proof.   To show the existence of U' and U'', it is sufficient by linearity and (3.76) to show this for D in $\mathfrak{D}(\mathfrak{C})$, R_s in $R_0(\mathfrak{C})$, L_t in $L_0(\mathfrak{C})$. If $U = D$ is in $\mathfrak{D}(\mathfrak{C})$, we may take $U' = U'' = D$. If $U = R_s$, we may take $U' = - R_s$,

$U'' = R_s + L_s$. For (3.90) is equivalent to $R_y U - U' R_y = R_{yU''}$, which holds by (3.69). Similarly, if $U = L_t$, we may take $U' = R_t + L_t$, $U'' = -L_t$. To show the uniqueness, it is sufficient to verify this for $U = 0$: if U', U'' are in $\mathfrak{o}(8, n)$, then

$$(3.91) \qquad (xU')y + x(yU'') = 0 \qquad \text{for all} \quad x, y \text{ in } \mathfrak{C}$$

implies $U' = U'' = 0$. Let $y = 1$ in (3.91); then $xU' = xR_g$ for g in $\mathfrak{C}$. Also $U' = R_g$ in $\mathfrak{o}(8, n)$ implies g is in $\mathfrak{C}_0$. Similarly, $x = 1$ in (3.91) implies $yU'' = yL_h$, h in $\mathfrak{C}$. Hence (3.91) becomes

$$(xg)y + x(hy) = 0 \qquad \text{for all} \quad x, y \text{ in } \mathfrak{C}.$$

Putting $x = y = 1$, we have $h = -g$, $(x, g, y) = 0$, g is in $F1 \cap \mathfrak{C}_0 = 0$. Hence $U' = U'' = 0$.

This theorem is sometimes called the "local" or "infinitesimal" principle of triality. The analogue for groups is proved in Jacobson [16, II]. Automorphisms of Cayley algebras are studied in Jacobson [14].

Let $\mathfrak{A}$ be a subalgebra of an algebra $\mathfrak{B}$. By a *derivation of* $\mathfrak{A}$ *into* $\mathfrak{B}$ is meant a linear transformation D of $\mathfrak{A}$ into $\mathfrak{B}$ satisfying

$$(xy)D = (xD)y + x(yD) \qquad \text{for all} \quad x, y \text{ in } \mathfrak{A}.$$

A generalization of Corollary 3.30 states that, if $\mathfrak{A}$ is a separable subalgebra of a finite-dimensional alternative algebra $\mathfrak{B}$ of characteristic $\neq 2, 3$, then any derivation D of $\mathfrak{A}$ into $\mathfrak{B}$ may be extended to an inner derivation of $\mathfrak{B}$. (Also characteristic 3 may be included in a fashion we shall neglect to mention.) This is equivalent to the analogue for alternative algebras of the first Whitehead lemma for Lie algebras (Jacobson [25], p. 77).

The *first Whitehead lemma for alternative algebras* (Taft [2]; Harris [2]) may be stated as follows: Let $\mathfrak{A}$ be a finite-dimensional separable alternative algebra over F of characteristic $\neq 2, 3$, let $\mathfrak{M}$ be a finite-dimensional alternative bimodule for $\mathfrak{A}$, and let $\mathfrak{B} = \mathfrak{A} + \mathfrak{M}$ be the split null extension. Let f be a one-cocycle of $\mathfrak{A}$ into $\mathfrak{M}$; that is, a linear transformation of $\mathfrak{A}$ into $\mathfrak{M}$ satisfying

$$f(ab) = f(a)b + af(b) \qquad \text{for all} \quad a, b \text{ in } \mathfrak{A}.$$

Then there exist g in the nucleus of $\mathfrak{B}$, x_i in $\mathfrak{A}$, and z_i in $\mathfrak{M}$, satisfying

$$f(a) = [a, g] + a \sum D_{x_i, z_i} \qquad \text{for all} \quad a \text{ in } \mathfrak{A},$$

where D_{x_i, z_i} is defined in (3.70). If F is of characteristic 0, we may take $g = 0$.

The first Whitehead lemma for Lie algebras (of characteristic 0) is used in the proof of the Theorem of Malcev-Harish-Chandra (Jacobson [25], p. 92). By a strictly analogous proof we have the following theorem: Let $\mathfrak{A}$ be a finite-dimensional alternative algebra of characteristic 0 with Wedderburn decomposition (3.49):

$$\mathfrak{A} = \mathfrak{S} + \mathfrak{N},$$

and let $\mathfrak{B}$ be a semisimple subalgebra of $\mathfrak{A}$. Then there exists a (nilpotent) derivation D of $\mathfrak{A}$ in the radical of the multiplication algebra $\mathfrak{M}(\mathfrak{A})$ such that the automorphism $G = \exp D$ of $\mathfrak{A}$ carries $\mathfrak{B}$ onto a subalgebra of $\mathfrak{S}$ (Schafer [8]). As a corollary we have the fact that, if $\mathfrak{A}$ has Wedderburn decompositions $\mathfrak{A} = \mathfrak{S} + \mathfrak{N} = \mathfrak{S}_1 + \mathfrak{N}$, then $\mathfrak{S}$ and $\mathfrak{S}_1$ are conjugate under an automorphism $G = \exp D$ of the type above. It would be interesting to have an extension of this corollary to arbitrary fields; the result is known in that generality for associative algebras (Malcev [1]).

As a final theorem in this chapter we mention the following: Let $\mathfrak{A}$ be a finite-dimensional alternative algebra of characteristic 0. Then $\mathfrak{A}$ is semisimple if and only if its derivation algebra $\mathfrak{D}(\mathfrak{A})$ is semisimple (or 0). The proof combines several concepts which we have developed separately in this chapter (Schafer [8]).

IV

JORDAN ALGEBRAS

1. THE RADICAL; SEMISIMPLE ALGEBRAS

In the Introduction we defined a (commutative) Jordan algebra $\mathfrak{J}$ over F to be a commutative algebra in which the *Jordan identity*

$$(4.1) \qquad (xy)x^2 = x(yx^2) \qquad \text{for all} \quad x, y \text{ in } \mathfrak{J}$$

is satisfied. Linearization of the Jordan identity requires that we assume F has characteristic $\neq 2$; we make this assumption throughout Chapter IV. It follows from (4.1) and the identities (4.2), (4.3) below that any scalar extension $\mathfrak{J}_K$ of a Jordan algebra $\mathfrak{J}$ is a Jordan algebra. Clearly any commutative associative algebra is a Jordan algebra.

Replacing x in

$$(x, y, x^2) = 0 \qquad \text{for all} \quad x, y \text{ in } \mathfrak{J}$$

by $x + \lambda z$ (λ in F), the coefficient of λ is 0 since F contains at least three distinct elements, and we have

$$(4.2) \qquad 2(x, y, zx) + (z, y, x^2) = 0 \qquad \text{for all} \quad x, y, z \text{ in } \mathfrak{J}.$$

Replacing x in (4.2) by $x + \lambda w$ ($\lambda \in F$), we have similarly (after dividing by 2) the multilinear identity

$$(4.3) \quad (x, y, wz) + (w, y, zx) + (z, y, xw) = 0 \qquad \text{for all} \quad w, x, y, z \text{ in } \mathfrak{J}.$$

91

Recalling that $L_a = R_a$ since $\mathfrak{J}$ is commutative, we see that (4.3) is equivalent to

$$(4.4) \quad [R_x, R_{wz}] + [R_w, R_{zx}] + [R_z, R_{xw}] = 0 \qquad \text{for all} \quad w, x, z \text{ in } \mathfrak{J}$$

and to

$$(4.5) \quad R_z R_{xy} - R_z R_y R_x + R_y R_{zx} - R_{y(zx)} + R_x R_{zy} - R_x R_y R_z = 0$$

for all x, y, z in $\mathfrak{J}$. Interchange x and y in (4.5) and subtract to obtain

$$(4.6) \quad [R_z, [R_x, R_y]] = R_{(x,z,y)} = R_{z[R_x, R_y]} \qquad \text{for all} \quad x, y, z \text{ in } \mathfrak{J}.$$

Now (4.6) says that, for all x, y in $\mathfrak{J}$, the operator $[R_x, R_y]$ is a derivation of $\mathfrak{J}$, since the defining condition for a derivation D of an arbitrary algebra $\mathfrak{A}$ may be written as

$$[R_z, D] = R_{zD} \qquad \text{for all} \quad z \text{ in } \mathfrak{A}.$$

Also (4.6) implies that the Lie multiplication algebra $\mathfrak{L}(\mathfrak{J})$ of $\mathfrak{J}$ is $\mathfrak{L}(\mathfrak{J}) = R(\mathfrak{J}) + [R(\mathfrak{J}), R(\mathfrak{J})]$. Hence the inner derivations of any Jordan algebra $\mathfrak{J}$ with unity element 1 are exactly the operators

$$D = \sum_i [R_{x_i}, R_{y_i}], \qquad x_i, y_i \text{ in } \mathfrak{J};$$

for, if $D = R_a + \sum [R_{x_i}, R_{y_i}]$ in $\mathfrak{L}(\mathfrak{J})$ is a derivation of $\mathfrak{J}$, then $0 = 1D = a + \sum (x_i, 1, y_i) = a$.

Our first objective is to prove that any Jordan algebra $\mathfrak{J}$ is power-associative. We define powers of x by $x^1 = x$, $x^{i+1} = xx^i$, and prove

$$(4.7) \qquad\qquad x^i x^j = x^{i+j} \qquad \text{for all} \quad x \text{ in } \mathfrak{J}.$$

For any x in $\mathfrak{J}$, write $\mathfrak{G}_x = \{R_x\} \cup \{R_{x^2}\}$. Then the enveloping algebra $\mathfrak{G}_x{}^*$ is commutative, since the generators R_x, R_{x^2} commute by (4.1). For $i \geq 2$, we put $y = x, z = x^{i-1}$ in (4.5) to obtain

$$(4.8) \qquad R_{x^{i+1}} = R_{x^{i-1}} R_{x^2} - R_{x^{i-1}} R_x{}^2 - R_x{}^2 R_{x^{i-1}} + 2R_x R_{x^i}.$$

By induction on i we see that R_{x^i} is in $\mathfrak{G}_x{}^*$ for $i = 3, 4, \dots$. Hence

$$R_{x^i} R_{x^j} = R_{x^j} R_{x^i} \qquad \text{for} \quad i, j = 1, 2, 3, \dots.$$

Then, in a proof of (4.7) by induction on i, we can assume that $x^i x^{j+1} = x^{i+j+1}$. It follows that

$$x^{i+1} x^j = (xx^i)x^j = xR_{x^i}R_{x^j} = xR_{x^j}R_{x^i} = x^{j+1}x^i = x^{i+j+1},$$

as desired. In the course of the proof we have seen that all R_{x^i} are contained in the subalgebra generated by R_x and R_{x^2}.

These are the only identities which we shall use directly. However, other important identities may be expressed in terms of

$$\{abc\} = (ab)c + (bc)a - (ac)b.$$

If $\mathfrak{A}$ is associative, then in $\mathfrak{A}^+$ we have $\{aba\} = 2(b \cdot a) \cdot a - b \cdot a^2 = aba$, so that $\{aba\}^2 = aba^2ba = \{a\{ba^2b\}a\}$. Hence

$$\{aba\}^2 = \{a\{ba^2b\}a\}$$

is satisfied in any special Jordan algebra. However, it has been proved that the free Jordan algebra with two generators is special (Shirshov [1]; Jacobson and Paige [1]). Therefore, this two-variable identity is true for arbitrary Jordan algebras. Also the identity

$$\{\{aba\}c\{aba\}\} = \{a\{b\{aca\}b\}a\}$$

is satisfied in any Jordan algebra (Macdonald [1]; Jacobson [20]). The class $\mathscr{C}$ of all homomorphic images of special Jordan algebras may be defined by identities (Cohn [3]); some identities which are valid in $\mathscr{C}$, but which do not hold for all Jordan algebras, are known (Glennie [1]).

We wish to prove the analogue for Jordan algebras of Theorem 3.2: any Jordan nilalgebra $\mathfrak{J}$ of finite dimension over F (of characteristic $\neq 2$) is nilpotent. The proof, due to Albert, is considerably more complicated than that of Theorem 3.2 because the identities involved in Jordan algebras are more complicated. However, the proof is not unduly long, and we include it in detail to provide a contrast with the simpler situation for alternative algebras. As in the proof of Theorem 3.2 we shall be concerned with a subalgebra $\mathfrak{B}$ of $\mathfrak{J}$ and elements x in $\mathfrak{J}$ satisfying (3.11): $x\mathfrak{B}^* \subseteq \mathfrak{B}$. Since $\mathfrak{J}$ is commutative, $x\mathfrak{B}^* \subseteq \mathfrak{B}$ is equivalent to $x\mathfrak{B} \subseteq \mathfrak{B}$.

Lemma 4.1. *Let $\mathfrak{B}$ be a subalgebra of a Jordan algebra $\mathfrak{J}$. If $x\mathfrak{B} \subseteq \mathfrak{B}$, then*

(i) $x^2\mathfrak{B}^2 \subseteq \mathfrak{B}$,

(ii) $(x^2\mathfrak{B})\mathfrak{B} \subseteq \mathfrak{B}$,

(iii) $(x^2\mathfrak{B})^2\mathfrak{B} \subseteq \mathfrak{B}$.

Proof. For nonempty subsets $\mathfrak{P}, \mathfrak{Q}, \mathfrak{R}$ of $\mathfrak{J}$, we use the notation $(\mathfrak{P}, \mathfrak{Q}, \mathfrak{R})$ for the *associator subspace* spanned by the associators (p, q, r), p in $\mathfrak{P}$, q in $\mathfrak{Q}$, r in $\mathfrak{R}$. Then (4.2) and (4.3) imply

$$(4.9) \qquad (x^2, \mathfrak{Q}, \mathfrak{P}) = (\mathfrak{P}, \mathfrak{Q}, x^2) \subseteq (x, \mathfrak{Q}, x\mathfrak{P}) = (x\mathfrak{P}, \mathfrak{Q}, x)$$

and

$$(4.10) \qquad\qquad (\mathfrak{P}, \mathfrak{Q}, \mathfrak{R}^2) \subseteq (\mathfrak{R}, \mathfrak{Q}, \mathfrak{P}\mathfrak{R}),$$

respectively. To prove (i), we compute

$$x^2\mathfrak{B}^2 \subseteq (x, x, \mathfrak{B}^2) + x(x\mathfrak{B}^2) \subseteq (\mathfrak{B}, x, x\mathfrak{B}) + x(x\mathfrak{B})$$
$$\subseteq (\mathfrak{B}, x, \mathfrak{B}) + x\mathfrak{B} \subseteq \mathfrak{B}$$

by (4.10). Then (4.9) and (i) imply (ii) as follows:

$$(x^2\mathfrak{B})\mathfrak{B} \subseteq (x^2, \mathfrak{B}, \mathfrak{B}) + x^2\mathfrak{B}^2 \subseteq (x\mathfrak{B}, \mathfrak{B}, x) + \mathfrak{B} \subseteq \mathfrak{B}.$$

For (iii) we begin with

$$(4.11) \qquad\qquad (x^2\mathfrak{B})^2\mathfrak{B} \subseteq [(x^2, \mathfrak{B}, x^2\mathfrak{B}) + x^2(\mathfrak{B}(x^2\mathfrak{B}))]\mathfrak{B}.$$

Now (4.9) implies

$$(x^2, \mathfrak{B}, x^2\mathfrak{B})\mathfrak{B} \subseteq (x(x^2\mathfrak{B}), \mathfrak{B}, x)\mathfrak{B} = (x^2(x\mathfrak{B}), \mathfrak{B}, x)\mathfrak{B}$$
$$\subseteq (x^2\mathfrak{B}, \mathfrak{B}, x)\mathfrak{B} \subseteq [((x^2\mathfrak{B})\mathfrak{B})x + (x^2\mathfrak{B})(\mathfrak{B}x)]\mathfrak{B}$$
$$\subseteq [\mathfrak{B}x + (x^2\mathfrak{B})\mathfrak{B}]\mathfrak{B} \subseteq \mathfrak{B}$$

by (ii). Also

$$[x^2(\mathfrak{B}(x^2\mathfrak{B}))]\mathfrak{B} \subseteq (x^2\mathfrak{B})\mathfrak{B} \subseteq \mathfrak{B}$$

by (ii). Hence (4.11) implies (iii).

Lemma 4.1 tells us that, if z is in $x^2\mathfrak{B}$, then $z\mathfrak{B} \subseteq \mathfrak{B}$ and $z^2\mathfrak{B} \subseteq \mathfrak{B}$. In the proof of Theorem 4.3 we shall also have occasion to use the fact that, if $z\mathfrak{B} \subseteq \mathfrak{B}$ and $z^2\mathfrak{B} \subseteq \mathfrak{B}$, z in $\mathfrak{J}$, then $z^k\mathfrak{B} \subseteq \mathfrak{B}$ for $k = 1, 2, 3, \ldots$. For we have seen that R_z and R_{z^2} generate all R_{z^k}. It follows that $z^k\mathfrak{B} \subseteq \mathfrak{B}$ for $k = 1, 2, 3, \ldots$, as desired.

Lemma 4.2. *Let $\mathfrak{B}$ be a finite-dimensional solvable subalgebra of a Jordan algebra $\mathfrak{J}$ (of characteristic $\neq 2$). Then $\mathfrak{B}^*$ is nilpotent.*

Proof. If $\mathfrak{B} \neq 0$, there exist a proper (solvable) subalgebra $\mathfrak{C}$ of $\mathfrak{B}$ and an element $w \notin \mathfrak{C}$ such that $\mathfrak{B} = Fw + \mathfrak{C}$ and $\mathfrak{B}^2 \subseteq \mathfrak{C}$. For $\mathfrak{B}/\mathfrak{B}^2$ ($\neq 0$) has a basis of cosets $w + \mathfrak{B}^2, w_2 + \mathfrak{B}^2, \ldots, w_m + \mathfrak{B}^2$. Then $\mathfrak{C} = Fw_2 + \cdots + Fw_m + \mathfrak{B}^2$ is a subalgebra of $\mathfrak{B}$. We use this fact to give a proof of the lemma by induction on the dimension of $\mathfrak{B}$. The result is clearly true if $\mathfrak{B} = 0$, and we assume that, for the solvable subalgebra $\mathfrak{C}$ above, $\mathfrak{C}^*$ is nilpotent: $(\mathfrak{C}^*)^k = 0$. Let

$$\mathfrak{H} = \mathfrak{B}^*\mathfrak{C}^* + \mathfrak{C}^*.$$

Since $\mathfrak{B}^2 \subseteq \mathfrak{C}$, we have

$$H_{x,y,z} = R_x R_{zy} + R_y R_{zx} + R_z R_{xy} - R_{y(zx)}$$

in $\mathfrak{H}$ for all x, y, z in $\mathfrak{B}$. Then (4.5) implies that

$$R_x R_y R_z + R_z R_y R_x = H_{x,y,z}$$

is in $\mathfrak{H}$ for all x, y, z in $\mathfrak{B}$. Every product of three right multiplications corresponding to elements of $\mathfrak{B} = Fw + \mathfrak{C}$ may be expressed as a linear combination of products $R_a R_b R_c$, where each of a, b, c is either in $\mathfrak{C}$ or equal to w. If c is in $\mathfrak{C}$, then $R_a R_b R_c$ is in $\mathfrak{B}^*\mathfrak{C}^* \subseteq \mathfrak{H}$. If $c = w$ and a is in $\mathfrak{C}$, then

$$R_a R_b R_c = H_{a,b,c} - R_c R_b R_a$$

is in $\mathfrak{H}$. Finally, if $a = c = w$, then

$$2R_a R_b R_c = H_{a,b,c} \in \mathfrak{H},$$

so $R_a R_b R_c$ is in $\mathfrak{H}$. Hence $R_x R_y R_z$ is in $\mathfrak{H}$ for all x, y, z in $\mathfrak{B}$, implying $(\mathfrak{B}^*)^3 \subseteq \mathfrak{H} = \mathfrak{B}^*\mathfrak{C}^* + \mathfrak{C}^*$. Then

$$(\mathfrak{B}^*)^4 \subseteq (\mathfrak{B}^*)^2\mathfrak{C}^* + \mathfrak{B}^*\mathfrak{C}^* = \mathfrak{B}^*\mathfrak{C}^*,$$

which is the case $i = 1$ of

(4.12) $$(\mathfrak{B}^*)^{3i+1} \subseteq \mathfrak{B}^*(\mathfrak{C}^*)^i, \qquad i = 1, 2, 3, \ldots.$$

Assuming (4.12), we have

$$(\mathfrak{B}^*)^{3i+4} \subseteq (\mathfrak{B}^*)^4(\mathfrak{C}^*)^i \subseteq \mathfrak{B}^*(\mathfrak{C}^*)^{i+1},$$

as desired. Then $(\mathfrak{C}^*)^k = 0$ implies $(\mathfrak{B}^*)^{3k+1} = 0$, $\mathfrak{B}^*$ is nilpotent.

Theorem 4.3 (Albert). *Any finite-dimensional Jordan nilalgebra* $\mathfrak{J}$ *(of characteristic* $\neq 2$) *is nilpotent.*

Proof. We prove first by induction on the dimension of $\mathfrak{J}$ that $\mathfrak{J}$ is solvable. If $\mathfrak{J}$ is generated by a single element, then the commutative associative algebra $\mathfrak{J}$ is nilpotent (hence solvable). Therefore, in our inductive proof, we may take a maximal proper subalgebra $\mathfrak{B} \neq 0$ of $\mathfrak{J}$ and know that $\mathfrak{B}$ is solvable. By Lemma 4.2, $\mathfrak{B}^*$ is nilpotent, $(\mathfrak{B}^*)^r = 0$ for some positive integer r. If $\mathfrak{J}\mathfrak{B} \subseteq \mathfrak{B}$, then $\mathfrak{B}$ is an ideal of $\mathfrak{J}$, and $\mathfrak{J}/\mathfrak{B}$ is a nilalgebra of lower dimension than $\mathfrak{J}$. Hence $\mathfrak{J}/\mathfrak{B}$ is solvable by the inductive hypothesis, and $\mathfrak{J}$ is solvable by Proposition 2.2. We assume therefore that $\mathfrak{B}$ does not contain $\mathfrak{J}\mathfrak{B}$. Then

$$\mathfrak{J}\mathfrak{B}^* \nsubseteq \mathfrak{B}, \qquad \mathfrak{J}(\mathfrak{B}^*)^r = 0 \subseteq \mathfrak{B},$$

imply that there exists a smallest integer $m \geq 2$ such that $\mathfrak{J}(\mathfrak{B}^*)^m \subseteq \mathfrak{B}$. Then $\mathfrak{B}$ does not contain $\mathfrak{J}(\mathfrak{B}^*)^{m-1}$, and there exists $x \in \mathfrak{J}(\mathfrak{B}^*)^{m-1}$, $x \notin \mathfrak{B}$, such that $x\mathfrak{B} \subseteq \mathfrak{B}$. If $\mathfrak{B}$ contains $x^2\mathfrak{B}$, we can apply the observation which precedes Lemma 4.2 to $z = x \notin \mathfrak{B}$. Otherwise, $\mathfrak{B}$ does not contain $x^2\mathfrak{B}$, and there exists b in $\mathfrak{B}$ such that $y = x^2b \notin \mathfrak{B}$. By Lemma 4.1 we have $y\mathfrak{B} \subseteq \mathfrak{B}$ and $y^2\mathfrak{B} \subseteq \mathfrak{B}$. Here we may apply the same observation to $z = y \notin \mathfrak{B}$. In both cases we have $z \notin \mathfrak{B}$ with $z^k\mathfrak{B} \subseteq \mathfrak{B}$ for $k = 1, 2, 3, \dots$. Since $\mathfrak{B}$ is a maximal proper subalgebra of $\mathfrak{J}$, the subalgebra of $\mathfrak{J}$ generated by $\mathfrak{B}$ and z is $\mathfrak{J}$ itself. But $z^k\mathfrak{B} \subseteq \mathfrak{B}$ for $k = 1, 2, 3, \dots$ implies that $\mathfrak{B}$ is an ideal of $\mathfrak{J}$, a contradiction. Hence $\mathfrak{J}$ is solvable. By Theorem 2.4 and Lemma 4.2, $\mathfrak{J}$ is nilpotent.

As in Chapter III, this means that there is a unique maximal nilpotent (= solvable = nil) ideal $\mathfrak{N}$ of any finite-dimensional Jordan algebra $\mathfrak{J}$ which is called the *radical* of $\mathfrak{J}$. Defining $\mathfrak{J}$ to be *semisimple* in case $\mathfrak{N} = 0$, we have seen that $\mathfrak{J}/\mathfrak{N}$ is semisimple. The proof that any semisimple Jordan algebra $\mathfrak{S}$ is a direct sum $\mathfrak{S} = \mathfrak{S}_1 \oplus \cdots \oplus \mathfrak{S}_t$ of simple $\mathfrak{S}_i$ is quite complicated for arbitrary F; we shall use a trace argument to give a proof by Dieudonné's Theorem 2.6 for F of characteristic 0.

As a consequence of Lemma 4.2, we have

Proposition 4.4. *Let* $\mathfrak{J}$ *be a (possibly infinite-dimensional) Jordan algebra of characteristic* $\neq 2$, *and* x *be a nilpotent element of* $\mathfrak{J}$. *Then* R_x *is nilpotent.*

Proof. Any nilpotent element x in $\mathfrak{J}$ generates a finite-dimensional nilpotent subalgebra $\mathfrak{B}$ of $\mathfrak{J}$. Lemma 4.2 implies that $\mathfrak{B}^*$ is nilpotent. Hence R_x in $\mathfrak{B}^*$ is nilpotent.

Let e be an idempotent in a Jordan algebra $\mathfrak{J}$. Put $i = 2$ and $x = e$ in (4.8) to obtain

$$2R_e{}^3 - 3R_e{}^2 + R_e = 0;$$

that is, $f(R_e) = 0$ where $f(\lambda) = (\lambda - 1)(2\lambda - 1)\lambda$. Hence the minimal polynomial for R_e divides $f(\lambda)$, and the only possibilities for characteristic roots of R_e are $1, \frac{1}{2}, 0$ (1 must occur since e is a characteristic vector belonging to the characteristic root $1 : eR_e = e^2 = e \neq 0$). Also the minimal polynomial for R_e has simple roots. Hence $\mathfrak{J}$ is the vector space direct sum

(4.13) $$\mathfrak{J} = \mathfrak{J}_1 + \mathfrak{J}_{1/2} + \mathfrak{J}_0,$$

where

$$\mathfrak{J}_i = \{x_i \,|\, x_i e = i x_i\}, \qquad i = 1, \tfrac{1}{2}, 0.$$

Taking a basis for $\mathfrak{J}$ adapted to the *Peirce decomposition* (4.13), we see that R_e has for its matrix relative to this basis the diagonal matrix $\mathrm{diag}\{1, 1, \ldots, 1, \frac{1}{2}, \frac{1}{2}, \ldots, \frac{1}{2}, 0, 0, \ldots, 0\}$ where the number of 1's is $\dim \mathfrak{J}_1 > 0$ and the number of $\frac{1}{2}$'s is $\dim \mathfrak{J}_{1/2}$. Hence

(4.14) $$\mathrm{trace}\, R_e = \dim \mathfrak{J}_1 + \tfrac{1}{2} \dim \mathfrak{J}_{1/2}.$$

If F has characteristic 0, then trace $R_e \neq 0$.

Theorem 4.5 (Albert). *The radical $\mathfrak{R}$ of any finite-dimensional Jordan algebra $\mathfrak{J}$ over F of characteristic 0 is the radical $\mathfrak{J}^\perp$ of the trace form*

(4.15) $$(x, y) = \mathrm{trace}\, R_{xy} \qquad \text{for all} \quad x, y \text{ in } \mathfrak{J}.$$

Proof. Without any assumption on the characteristic of F it follows from (4.6) that (x, y) is a trace form:

$$(xy, z) - (x, yz) = \mathrm{trace}\, R_{(x,y,z)} = \mathrm{trace}\,[R_z, [R_x, R_y]] = 0$$

since the trace of any commutator is 0. Hence $\mathfrak{J}^\perp$ is an ideal of $\mathfrak{J}$. If $\mathfrak{J}^\perp$ were not a nilideal, then (by Proposition 3.3) $\mathfrak{J}^\perp$ would contain an idempotent e ($\neq 0$) and, assuming characteristic 0, $(e, e) = \mathrm{trace}\, R_e \neq 0$ by (4.14), a contradiction. Hence $\mathfrak{J}^\perp$ is a nilideal and $\mathfrak{J}^\perp \subseteq \mathfrak{R}$.

Conversely, if x is in $\mathfrak{N}$, then xy is in $\mathfrak{N}$ for every y in $\mathfrak{J}$, and R_{xy} is nilpotent by Proposition 4.4. Hence $(x, y) = \text{trace } R_{xy} = 0$ for all y in $\mathfrak{J}$; that is, x is in $\mathfrak{J}^{\perp}$. Hence $\mathfrak{N} \subseteq \mathfrak{J}^{\perp}$, $\mathfrak{N} = \mathfrak{J}^{\perp}$.

Corollary 4.6. *Any (finite-dimensional) semisimple Jordan algebra $\mathfrak{J}$ over F of characteristic 0 is (uniquely) expressible as a direct sum $\mathfrak{J} = \mathfrak{S}_1 \oplus \cdots \oplus \mathfrak{S}_t$ of simple ideals $\mathfrak{S}_i$.*

Proof. By Theorem 4.5 the (associative) trace form $(x, y) = \text{trace } R_{xy}$ is nondegenerate; hence hypothesis (i) in Theorem 2.6 is satisfied. Also any ideal $\mathfrak{B}$ such that $\mathfrak{B}^2 = 0$ is nilpotent; hence $\mathfrak{B} = 0$, establishing (ii).

As mentioned above, Corollary 4.6 is actually true for F of characteristic $\neq 2$ (Albert [13]; Jacobson [23]). What remains then, as far as the structure of semisimple Jordan algebras is concerned, is a determination of the central simple algebras. The first step in this is to show that every semisimple $\mathfrak{J}$ (hence every simple $\mathfrak{J}$) has a unity element 1. Again the argument we use here is valid only for characteristic 0, whereas the theorem is true in general (Albert [13]; Jacobson [23]).

We begin by returning to the Peirce decomposition (4.13) of any Jordan algebra $\mathfrak{J}$ relative to an idempotent e. The subspaces $\mathfrak{J}_1$ and $\mathfrak{J}_0$ are orthogonal subalgebras of $\mathfrak{J}$ which are related to the subspace $\mathfrak{J}_{1/2}$ as follows:

$$(4.16) \quad \mathfrak{J}_{1/2}\mathfrak{J}_{1/2} \subseteq \mathfrak{J}_1 + \mathfrak{J}_0, \qquad \mathfrak{J}_1 \mathfrak{J}_{1/2} \subseteq \mathfrak{J}_{1/2}, \qquad \mathfrak{J}_0 \mathfrak{J}_{1/2} \subseteq \mathfrak{J}_{1/2}.$$

For we may put $x = e$, $z = x_i \in \mathfrak{J}_i$, $y = y_j \in \mathfrak{J}_j$ in (4.2) to obtain $2i(e, y_j, x_i) + (x_i, y_j, e) = 0$, or $(1 - 2i)[(x_i y_j)e - j(x_i y_j)] = 0$, so that

$$\mathfrak{J}_i \mathfrak{J}_j \subseteq \mathfrak{J}_j \qquad \text{if} \quad i \neq \tfrac{1}{2}.$$

Hence

$$\mathfrak{J}_1{}^2 \subseteq \mathfrak{J}_1, \qquad \mathfrak{J}_0{}^2 \subseteq \mathfrak{J}_0, \qquad \mathfrak{J}_1 \mathfrak{J}_0 = \mathfrak{J}_0 \mathfrak{J}_1 \subseteq \mathfrak{J}_0 \cap \mathfrak{J}_1 = 0,$$

so $\mathfrak{J}_1$ and $\mathfrak{J}_0$ are orthogonal subalgebras, and also the last two inclusions in (4.16) hold. Put $x = x_{1/2}$, $z = y_{1/2}$, $y = w = e$ in (4.3) and write $x_{1/2}y_{1/2} = a = a_1 + a_{1/2} + a_0$ to obtain

$$\tfrac{1}{2}(x_{1/2}, e, y_{1/2}) + (e, e, a) + \tfrac{1}{2}(y_{1/2}, e, x_{1/2}) = (e, e, a) = 0.$$

Hence

$$ea - e(ea) = a_1 + \tfrac{1}{2}a_{1/2} - e(a_1 + \tfrac{1}{2}a_{1/2})$$
$$= a_1 + \tfrac{1}{2}a_{1/2} - a_1 - \tfrac{1}{4}a_{1/2} = \tfrac{1}{4}a_{1/2} = 0.$$

Hence $x_{1/2}y_{1/2} = a_1 + a_0$ in $\mathfrak{J}_1 + \mathfrak{J}_0$, establishing (4.16).

Now

(4.17) $\qquad\qquad$ trace $R_b = 0 \qquad$ for all $\quad b$ in $\mathfrak{J}_{1/2}$.

For b in $\mathfrak{J}_{1/2}$ implies

$$\text{trace } R_b = 2 \text{ trace } R_{eb} = 2(e, b) = 2(e^2, b) = 2(e, eb) = (e, b)$$

since (4.15) is a trace form, so trace $R_b = 0$. If $x = x_1 + x_{1/2} + x_0$, $y = y_1 + y_{1/2} + y_0$ in accordance with (4.13), we have

$$xy = (x_1 y_1 + x_{1/2} y_{1/2} + x_0 y_0) + (x_1 y_{1/2} + x_{1/2} y_1 + x_{1/2} y_0 + x_0 y_{1/2})$$

with the last term in parentheses in $\mathfrak{J}_{1/2}$ by (4.16). Hence (4.17) implies that

$$(x, y) = \text{trace } R_{x_1 y_1 + x_{1/2} y_{1/2} + x_0 y_0}.$$

Now $x_{1/2} y_{1/2} = c = c_1 + c_0$ (c_i in $\mathfrak{J}_i$) implies

$$\text{trace } R_{c_1} + \text{trace } R_{c_0} = \text{trace } R_c = (x_{1/2}, y_{1/2}) = 2(ex_{1/2}, y_{1/2})$$
$$= 2(e, x_{1/2} y_{1/2}) = 2 \text{ trace } R_{e(c_1 + c_0)}$$
$$= 2 \text{ trace } R_{c_1},$$

so that trace $R_{c_1} = \text{trace } R_{c_0}$. Hence

(4.18) $\qquad (x, y) = \text{trace } R_{x_1 y_1 + z_0}, \qquad z_0 = 2c_0 + x_0 y_0 \quad \text{in } \mathfrak{J}_0.$

Just as for alternative algebras, it is easy to see that any finite-dimensional Jordan algebra $\mathfrak{J}$ which is not a nilalgebra contains an idempotent e with the property that there is no idempotent in the subalgebra $\mathfrak{J}_0$ given by the Peirce decomposition (4.13) relative to e; equivalently, $\mathfrak{J}_0$ is a nilalgebra. Such an idempotent in $\mathfrak{J}$ is a *principal* idempotent.

Theorem 4.7. *Any semisimple Jordan algebra $\mathfrak{J} \neq 0$ of finite dimension over F of characteristic 0 has a unity element 1.*

Proof. $\mathfrak{J}$ has a principal idempotent e. Then $\mathfrak{J}_0$ is a nilalgebra, so that $(x, y) = $ trace $R_{x_1 y_1}$ by (4.18) and Proposition 4.4. Hence x in $\mathfrak{J}_{1/2} + \mathfrak{J}_0$ implies $x_1 = 0$, $(x, y) = 0$ for all y in $\mathfrak{J}$, so x is in $\mathfrak{J}^\perp$. That is, $\mathfrak{J}_{1/2} + \mathfrak{J}_0 \subseteq \mathfrak{J}^\perp = \mathfrak{N} = 0$, or $\mathfrak{J} = \mathfrak{J}_1$, $e = 1$.

If $\mathfrak{J}$ contains 1 and $e_1 \neq 1$, then $e_2 = 1 - e_1$ is an idempotent, and the Peirce decompositions relative to e_1 and e_2 coincide (with differing subscripts). We introduce a new notation: $\mathfrak{J}_{11} = \mathfrak{J}_{1,e_1}$ $(= \mathfrak{J}_{0,e_2})$, $\mathfrak{J}_{12} = \mathfrak{J}_{1/2,e_1}$ $(=\mathfrak{J}_{1/2,e_2})$, $\mathfrak{J}_{22} = \mathfrak{J}_{0,e_1}$ $(= \mathfrak{J}_{1, e_2})$. More generally, if $1 = e_1 + e_2 + \cdots + e_t$ for pairwise orthogonal idempotents e_i, we have the refined Peirce decomposition

$$(4.19) \qquad\qquad \mathfrak{J} = \sum_{i \leq j} \mathfrak{J}_{ij}$$

of $\mathfrak{J}$ as the vector space direct sum of subspaces

$$\mathfrak{J}_{ii} = \mathfrak{J}_{1,e_i} \qquad\qquad (1 \leq i \leq t),$$

$$\mathfrak{J}_{ij} = \mathfrak{J}_{1/2,e_i} \cap \mathfrak{J}_{1/2,e_j} \qquad (1 \leq i < j \leq t);$$

that is,

$$(4.20) \qquad \mathfrak{J}_{ii} = \{x \mid x \in \mathfrak{J}, \quad xe_i = x\},$$

$$\mathfrak{J}_{ij} = \mathfrak{J}_{ji} = \{x \mid x \in \mathfrak{J}, \quad xe_i = \tfrac{1}{2}x = xe_j\}, \qquad i \neq j.$$

Multiplicative relationships among the $\mathfrak{J}_{ij}$ are consequences of (4.16) and the statement preceding it.

2. CENTRAL SIMPLE JORDAN ALGEBRAS

We recall that an idempotent e in $\mathfrak{J}$ is called *primitive* in case e cannot be written as the sum $e = u + v$ of orthogonal idempotents; that is, e is the only idempotent in $\mathfrak{J}_1$. Also e is *absolutely primitive* in case it is primitive in any scalar extension $\mathfrak{J}_K$ of $\mathfrak{J}$. A finite-dimensional central simple Jordan algebra $\mathfrak{J}$ is called *reduced* in case $1 = e_1 + \cdots + e_t$ for pairwise orthogonal absolutely primitive idempotents e_i in $\mathfrak{J}$. In this case it can be shown that the subalgebras $\mathfrak{J}_{ii}$ in the Peirce decomposition (4.20) are 1-dimensional ($\mathfrak{J}_{ii} = Fe_i$). If $\mathfrak{J}$ is a finite-dimensional central simple algebra over F, there is a scalar extension $\mathfrak{J}_K$ which is reduced (for example, take K to be the algebraic closure of F), and it can be shown that the number t of pairwise orthogonal absolutely primitive idempotents e_i in $\mathfrak{J}_K$ such that $1 = e_1 + \cdots + e_t$ is unique; t is called the *degree* of $\mathfrak{J}$.

We list without proof all (finite-dimensional) central simple Jordan algebras $\mathfrak{J}$ of degree t over F of characteristic $\neq 2$ (Albert [6], [13]; Kalisch [1]; Jacobson and Jacobson [1]; Jacobson [13]). Recall from the Introduction that $\mathfrak{J}$ is a *special Jordan algebra* in case $\mathfrak{J}$ is isomorphic to a subalgebra of an algebra $\mathfrak{A}^+$ where $\mathfrak{A}$ is associative and multiplication in $\mathfrak{A}^+$ is defined by

(4.21) $$x \cdot y = \tfrac{1}{2}(xy + yx).$$

We say that each algebra is of *type* A, B, C, D, or E listed below.

A_I. $\mathfrak{J} \cong \mathfrak{A}^+$ with $\mathfrak{A}$ any central simple associative algebra (necessarily of dimension t^2 over F).

A_{II}. Let $\mathfrak{A}$ be any involutorial simple associative algebra over F, the involution being of the second kind (so that the center $\mathfrak{Z}$ of $\mathfrak{A}$ is a quadratic extension of F and the involution induces a nontrivial automorphism on $\mathfrak{Z}$) (Albert [24], p. 153). Then $\mathfrak{J} \cong \mathfrak{H}(\mathfrak{A})$, the t^2-dimensional subalgebra of self-adjoint elements in the $2t^2$-dimensional algebra $\mathfrak{A}^+$. If $\mathfrak{J}$ is of type A_I or A_{II}, and if K is the algebraic closure of F, then $\mathfrak{J}_K \cong K_t^+$ where K_t is the algebra of all $t \times t$ matrices with elements in K.

B, C. Let $\mathfrak{A}$ be any involutorial central simple associative algebra over F (so the involution is of the first kind). Then $\mathfrak{J} \cong \mathfrak{H}(\mathfrak{A})$, the subalgebra of self-adjoint elements in $\mathfrak{A}^+$. There are two types (B and C) which may be distinguished by passing to the algebraic closure K of F, so that $\mathfrak{A}_K$ is a total matrix algebra. In case B the (extended) involution on $\mathfrak{A}_K$ is transposition ($a \to a'$) so that $\mathfrak{A}$ has dimension t^2 and $\mathfrak{J}$ has dimension $\tfrac{1}{2}t(t+1)$ over F. In case C the (extended) involution on $\mathfrak{A}_K$ is $a \to g^{-1}a'g$ where $g = \begin{pmatrix} 0 & 1_t \\ -1_t & 0 \end{pmatrix}$ so that $\mathfrak{A}$ has dimension $4t^2$ and $\mathfrak{J}$ has dimension $2t^2 - t$ over F.

D. Let (x, y) be any nondegenerate symmetric bilinear form on a vector space $\mathfrak{M}$ of dimension $n \geq 2$. Then $\mathfrak{J}$ is the vector space direct sum $\mathfrak{J} = F1 + \mathfrak{M}$, multiplication in the $(n+1)$-dimensional algebra $\mathfrak{J}$ being defined by $xy = (x, y)1$ for all x, y in $\mathfrak{M}$. Here $t = 2$ ($\dim \mathfrak{J} \geq 3$).

E.   The algebra $\mathfrak{C}_3$ of all 3×3 matrices with elements in a Cayley algebra $\mathfrak{C}$ over F has the *standard involution* $x \to \bar{x}'$ (conjugate transpose). The 27-dimensional subspace $\mathfrak{H}(\mathfrak{C}_3)$ of self-adjoint elements

$$(4.22) \qquad x = \begin{pmatrix} \xi_1 & c & \bar{b} \\ \bar{c} & \xi_2 & a \\ b & \bar{a} & \xi_3 \end{pmatrix}, \qquad \xi_i \text{ in } F, \quad a, b, c \text{ in } \mathfrak{C},$$

is a (central simple) Jordan algebra of degree $t = 3$ under the multiplication (4.21) where xy is the multiplication in $\mathfrak{C}_3$ (which is not associative). Then $\mathfrak{J}$ is any algebra such that some scalar extension $\mathfrak{J}_K \cong \mathfrak{H}(\mathfrak{C}_3)_K \,(= \mathfrak{H}((\mathfrak{C}_K)_3))$.

A central simple Jordan algebra of degree 2 (that is, of type D) is a commutative quadratic algebra with 1 $(a^2 - t(a)a + n(a)1 = 0)$ having nondegenerate norm form $n(a)$, and conversely. For $a = \alpha 1 + x$, $x \in \mathfrak{M}$, implies $a^2 - t(a)a + n(a)1 = 0$, where $t(a) = 2\alpha$, $n(a) = \alpha^2 - (x, x)$, and $n(a)$ is nondegenerate if and only if (x, y) is.

The algebras of types A, B, C are special Jordan algebras by definition. An algebra of type D is a subalgebra of $\mathfrak{A}^+$, where $\mathfrak{A}$ is the (associative) Clifford algebra of (x, y) (Artin [2], p. 186). But algebras of type E are not special (as we show below), and are therefore called *exceptional* central simple Jordan algebras. Exceptional Jordan division algebras exist over suitable fields F; but not, for example, over a finite field or the field of all real numbers (Albert [20]). If an exceptional central simple Jordan algebra $\mathfrak{J}$ is not a division algebra, then it is reduced, and $\mathfrak{J}$ is isomorphic to an algebra $\mathfrak{H}(\mathfrak{C}_3, \Gamma)$ of self-adjoint elements in $\mathfrak{C}_3$ under a *canonical involution* $x \to \Gamma^{-1}\bar{x}'\Gamma$ where $\Gamma = \text{diag}\{\gamma_1, \gamma_2, \gamma_3\}$, $\gamma_i \neq 0$ in F. Isomorphism of reduced exceptional simple Jordan algebras is studied in Albert and Jacobson [1].

The unifying feature in the list of central simple Jordan algebras above is that, for $t > 2$, a reduced central simple Jordan algebra is isomorphic to the algebra $\mathfrak{H}(\mathfrak{D}_t, \Gamma)$ defined as follows: $\mathfrak{D}$ is a composition algebra (that is, an alternative algebra of dimension 1, 2, 4, or 8 with unity element u and involution $d \to \bar{d}$ satisfying $d + \bar{d} \in Fu$, $d\bar{d} = n(d)u$, $n(d)$ nondegenerate on $\mathfrak{D}$); $\mathfrak{D}_t$ is the algebra of all $t \times t$ matrices with elements in $\mathfrak{D}$; $\Gamma = \text{diag}\{\gamma_1, \gamma_2, \ldots, \gamma_t\}$, $\gamma_i \neq 0$ in F. Then $x \to \Gamma^{-1}\bar{x}'\Gamma$ is a canonical involution in $\mathfrak{D}_t$, and the set $\mathfrak{H}(\mathfrak{D}_t, \Gamma)$ of all self-adjoint elements in $\mathfrak{D}_t$ is a subalgebra of $\mathfrak{D}_t^+$; that is, we do not need $\mathfrak{A}$ associative to define $\mathfrak{A}^+$ by (4.21). If $\mathfrak{D}$ is associative, then

$\mathfrak{D}_t = \mathfrak{D} \otimes_F F_t$ is associative, and $\mathfrak{J} \cong \mathfrak{H}(\mathfrak{D}_t, \Gamma)$ is a special Jordan algebra. If $\mathfrak{D}$ is not associative, then $\mathfrak{J} \cong \mathfrak{H}(\mathfrak{D}_t, \Gamma)$ is not a Jordan algebra unless $t = 3$. Hence we have $\mathfrak{J}$ of type B if $\mathfrak{D} = F1$; $\mathfrak{J}$ of type A if $\mathfrak{D} = \mathfrak{Z}$ (type A_I if $\mathfrak{Z} = F \oplus F$; type A_{II} if $\mathfrak{Z}$ is a quadratic field over F); $\mathfrak{J}$ of type C if $\mathfrak{D} = \mathfrak{Q}$; $\mathfrak{J}$ of type E if $t = 3$ and $\mathfrak{D} = \mathfrak{C}$. The corresponding dimensions for $\mathfrak{J}$ are clearly $t + \frac{1}{2}t(t-1)(\dim \mathfrak{D})$; that is, $\frac{1}{2}t(t+1)$ for type B, t^2 for type A, $2t^2 - t$ for type C, and 27 for type E.

We have stated these results without proof, but it is interesting to have at least an indication of how this relationship between alternative and Jordan algebras emerges. Let $\mathfrak{J}$ be a reduced central simple Jordan algebra of degree $t > 2$ over F. Then $\mathfrak{J}_{ii} = Fe_i \, (i = 1, \dots, t)$ in the Peirce decomposition (4.19). Since $\mathfrak{J}_{ij}\mathfrak{J}_{ij} \subseteq \mathfrak{J}_{ii} + \mathfrak{J}_{jj} = Fe_i + Fe_j$, we have $x_{ij}^2 = 4\alpha e_i + 4\beta e_j$ for all x_{ij} in $\mathfrak{J}_{ij}$ $(i \neq j)$. Then $(x_{ij}^2 e_i)x_{ij} = x_{ij}^2(e_i x_{ij})$ yields $2\alpha x_{ij} = (\alpha + \beta)x_{ij}$, or $\alpha = \beta$. That is,

$$x_{ij}^2 = 4N_{ij}(x_{ij})(e_i + e_j) \qquad \text{for all} \quad x_{ij} \text{ in } \mathfrak{J}_{ij},$$

where $N_{ij}(x_{ij})$ is a quadratic form on $\mathfrak{J}_{ij}$, which may be shown to be nondegenerate. For i, j, k distinct, the properties of the Peirce decomposition include

$$\mathfrak{J}_{ij}\mathfrak{J}_{jk} \subseteq \mathfrak{J}_{ki} \qquad (i, j, k \text{ distinct});$$

put $x = y_{jk}, y = e_i, z = x_{ij}$ in (4.2) to obtain

$$(4.23) \qquad (x_{ij}y_{jk})y_{jk} = N_{jk}(y_{jk})x_{ij} \qquad \text{for all} \quad x_{ij} \text{ in } \mathfrak{J}_{ij}, \quad y_{jk} \text{ in } \mathfrak{J}_{jk}.$$

Next put $x = x_{ij}, y = z = y_{jk}$ in (4.2) to obtain

$$(4.24) \quad N_{ki}(x_{ij}y_{jk}) = N_{ij}(x_{ij})N_{jk}(y_{jk}) \qquad \text{for all} \quad x_{ij} \text{ in } \mathfrak{J}_{ij}, \quad y_{jk} \text{ in } \mathfrak{J}_{jk}$$

$(i, j, k$ distinct). Take u_{jk} in $\mathfrak{J}_{jk}$ with $N_{jk}(u_{jk}) \neq 0$; it is convenient to normalize this choice so that

$$N_{23}(u_{23})N_{31}(u_{31}) = 1.$$

Equation (4.24) and the nondegeneracy of $N_{ij}(x_{ij})$ imply that the linear mapping $x_{ij} \to x_{ij}u_{jk}$ is (1-1) from $\mathfrak{J}_{ij}$ into $\mathfrak{J}_{ki}$. Hence $\dim \mathfrak{J}_{ij} \leq \dim \mathfrak{J}_{ki} \leq \dim \mathfrak{J}_{jk} \leq \dim \mathfrak{J}_{ij}$ implies that the subspaces $\mathfrak{J}_{ij} \, (i \neq j)$ all have the same dimension and that $x_{ij} \to x_{ij}u_{jk}$ is a vector space isomorphism between $\mathfrak{J}_{ij}$ and $\mathfrak{J}_{ki}$ (i, j, k distinct). Let d and d' be arbitrary elements in $\mathfrak{J}_{12}$. Then $(u_{31}d)(d'u_{23})$ is in $\mathfrak{J}_{23}\mathfrak{J}_{31} \subseteq \mathfrak{J}_{12}$, so that

$$d * d' = (u_{31}d)(d'u_{23}) \qquad \text{for all} \quad d, d' \text{ in } \mathfrak{J}_{12}$$

is a bilinear multiplication in $\mathfrak{J}_{12}$. We write $\mathfrak{D}$ for the algebra $\mathfrak{J}_{12}$ equipped with this multiplication, and $N(d)$ for the nondegenerate quadratic form $N_{12}(d)$ on $\mathfrak{D}$. Now $u = u_{23}u_{31}$ in $\mathfrak{J}_{12}$ is a unity element in $\mathfrak{D}$, since

$$d * u = (u_{31}d)((u_{23}u_{31})u_{23}) = N_{23}(u_{23})((u_{31}d)u_{31})$$

$$= N_{23}(u_{23})N_{31}(u_{31})d = d$$

by (4.23), and similarly $u * d = d$ for all d in $\mathfrak{D}$. Also

$$N(d * d') = N_{12}((u_{31}d)(d'u_{23})) = N_{23}(u_{31}d)N_{31}(d'u_{23})$$

$$= N_{31}(u_{31})N_{12}(d)N_{12}(d')N_{23}(u_{23}) = N(d)N(d')$$

for all d, d' in $\mathfrak{D}$ by (4.24); that is, $N(d)$ permits composition. Hurwitz's Theorem 3.25 implies that $\mathfrak{D}$ is a composition algebra, one of the alternative algebras of dimension 1, 2, 4 or 8 studied in Chapter III.

Theorem 4.8 (Albert). *Any central simple Jordan algebra $\mathfrak{J}$ of type* E *is exceptional* (*that is, is not a special Jordan algebra*).

Proof. It is sufficient to prove that $\mathfrak{H}(\mathbb{C}_3)$ is not special. For, if $\mathfrak{J}$ were special, then $\mathfrak{J} \cong \mathfrak{J}' \subseteq \mathfrak{A}^+$ with $\mathfrak{A}$ associative implies $\mathfrak{J}_K = K \otimes \mathfrak{J} \cong K \otimes \mathfrak{J}' \subseteq K \otimes \mathfrak{A}^+ = (K \otimes \mathfrak{A})^+ = \mathfrak{A}_K{}^+$ so that $\mathfrak{H}((\mathbb{C}_K)_3) \cong \mathfrak{J}_K$ is special, a contradiction.

Suppose that $\mathfrak{H}(\mathbb{C}_3)$ is special. There is an associative algebra $\mathfrak{A}$ (of possibly infinite dimension over F) such that U is an isomorphism of $\mathfrak{H}(\mathbb{C}_3)$ into $\mathfrak{A}^+$. For $i = 1, 2, 3$ define elements e_i in $\mathfrak{A}$ and 8-dimensional subspaces

$$\mathfrak{S}_i = \{d_i \mid d \in \mathbb{C}\}$$

of $\mathfrak{A}$ by

$$xU = \xi_1 e_1 + \xi_2 e_2 + \xi_3 e_3 + a_1 + b_2 + c_3$$

for x in (4.22); that is, for ξ_i in F and a, b, c in $\mathfrak{C}$. (Note that our notation is such that we will never use e for an element of $\mathfrak{C}$.) Then

$$\mathfrak{S} = Fe_1 + Fe_2 + Fe_3 + \mathfrak{S}_1 + \mathfrak{S}_2 + \mathfrak{S}_3 = \mathfrak{H}(\mathfrak{C}_3)U$$

is a 27-dimensional subspace of $\mathfrak{A}$. $\mathfrak{S}$ is a subalgebra of $\mathfrak{A}^+$. The mapping $V = U^{-1}$ defined on $\mathfrak{S}$ (not on all of $\mathfrak{A}$) is an isomorphism of $\mathfrak{S}$ onto $\mathfrak{H}(\mathfrak{C}_3)$:

$$(4.25) \qquad (xU \cdot yU)V = x \cdot y \qquad \text{for all} \quad x, y \text{ in } \mathfrak{H}(\mathfrak{C}_3).$$

Performing the multiplications in $\mathfrak{H}(\mathfrak{C}_3)$, we see that (4.25) yields

$$(4.26) \qquad e_i^2 = e_i \quad (\neq 0), \qquad i = 1, 2, 3;$$

$$(4.27) \quad e_i \cdot e_j = 0, \qquad i \neq j;$$

$$(4.28) \quad e_i \cdot a_i = 0, \qquad a \text{ in } \mathfrak{C}, \quad i = 1, 2, 3;$$

$$(4.29) \quad e_i \cdot a_j = \tfrac{1}{2}a_j, \qquad a \text{ in } \mathfrak{C}, \quad i \neq j;$$

$$(4.30) \quad a_i \cdot b_i = (a, b)(e_j + e_k), \qquad a, b \text{ in } \mathfrak{C}, \quad i, j, k \text{ distinct,}$$

where (a, b) is the nondegenerate symmetric bilinear form (3.60) associated with $n(a)$; and

$$(4.31) \quad 2a_i \cdot b_j = (\bar{b}\bar{a})_k, \qquad a, b \text{ in } \mathfrak{C}, \quad i, j, k \text{ a cyclic permutation of } 1, 2, 3.$$

Now (4.26) and (4.27) imply that e_i $(i = 1, 2, 3)$ are pairwise orthogonal idempotents. For $\mathfrak{A}$ is associative, so $e_i e_j + e_j e_i = 0$ for $i \neq j$ implies

$$e_i^2 e_j + e_i e_j e_i = 0 = e_i e_j e_i + e_j e_i^2,$$

or $e_i e_j = e_j e_i$; hence $e_i e_j = 0$ for $i \neq j$. By an identical proof it follows from (4.28) that

$$(4.32) \qquad e_i a_i = a_i e_i = 0, \qquad i = 1, 2, 3.$$

For i, j, k distinct, (4.29) implies $e_i a_j + a_j e_i = a_j = e_k a_j + a_j e_k$; then $f a_j + a_j f = 2a_j$ for the idempotent $f = e_i + e_k$. Hence $f^2 a_j + f a_j f = 2f a_j$, so $f a_j f = f a_j$, and similarly $f a_j f = a_j f$; that is, $f a_j = a_j f = a_j$:

$$(4.33) \qquad (e_i + e_k)a_j = a_j = a_j(e_i + e_k), \qquad i, j, k \text{ distinct.}$$

Also (4.29) implies $e_i a_j = a_j - a_j e_i$, so $e_i a_j e_i = a_j e_i - a_j e_i^2 = 0$:

(4.34) $$e_i a_j e_i = 0, \qquad i \neq j.$$

For any a in $\mathfrak{C}$, define

$$a' = e_1 a_3 u_3 \qquad \text{in } \mathfrak{A},$$

where u is the unity element of $\mathfrak{C}$. Then

$$(ab)' = e_1(ab)_3 u_3 = e_1(\bar{b}_1 \bar{a}_2 + \bar{a}_2 \bar{b}_1)u_3 = e_1 \bar{a}_2 \bar{b}_1 u_3$$

by (4.31) and (4.32). Also (4.31) implies $a_3 u_1 + u_1 a_3 = (\bar{u}\bar{a})_2 = \bar{a}_2$ and

(4.35) $$u_2 b_3 + b_3 u_2 = \bar{b}_1.$$

Hence $(ab)' = e_1(a_3 u_1 + u_1 a_3)(u_2 b_3 + b_3 u_2)u_3 = e_1 a_3 u_1(u_2 b_3 + b_3 u_2)u_3$ by (4.32). Now

$$b_3 u_2 u_3 = b_3 u_2(e_1 + e_3)u_3 = b_3 u_2 e_1 u_3 = (\bar{b}_1 - u_2 b_3)e_1 u_3 = -u_2 b_3 e_1 u_3$$
$$= -u_2(e_1 + e_3)b_3 e_1 u_3 = -u_2 e_1 b_3 e_1 u_3 = 0$$

by (4.32)–(4.35). Also

$$u_1 u_2 u_3 = u_1 u_2(e_1 + e_2)b_3 = u_1 u_2 e_1 b_3 = (u_3 - u_2 u_1)e_1 b_3 = u_3 e_1 b_3.$$

Hence

$$(ab)' = e_1 a_3 u_1 u_2 b_3 u_3 = e_1 a_3 u_3 e_1 b_3 u_3 = a'b'.$$

Clearly the mapping $a \to a'$ is linear; hence it is a homomorphism of $\mathfrak{C}$ onto the subalgebra $\mathfrak{C}'$ of $\mathfrak{A}$ consisting of all a'. Since $\mathfrak{C}$ is simple, the kernel of this homomorphism is either 0 or $\mathfrak{C}$; in the latter case $0 = u' = e_1 u_3^2 = e_1(e_1 + e_2) = e_1 \neq 0$ by (4.30), and we have a contradiction. Hence $a \to a'$ is an isomorphism. But $\mathfrak{C}'$ is associative, whereas $\mathfrak{C}$ is not. Hence $\mathfrak{H}(\mathfrak{C}_3)$ is an exceptional Jordan algebra.

Just as for alternative algebras, a finite-dimensional Jordan algebra $\mathfrak{J}$ is separable if and only if $\mathfrak{J} = \mathfrak{S}_1 \oplus \cdots \oplus \mathfrak{S}_t$ is semisimple with the center $\mathfrak{C}_i$ of each simple component $\mathfrak{S}_i$ being a separable extension of F.

We state without proof the *Wedderburn principal theorem for Jordan algebras* which is analogous to Theorem 3.18: Let $\mathfrak{J}$ be a finite-dimensional Jordan algebra over F of characteristic $\neq 2$, and let $\mathfrak{N}$ be its radical. If $\mathfrak{J}/\mathfrak{N}$ is separable, then $\mathfrak{J} = \mathfrak{S} + \mathfrak{N}$ (direct sum) where $\mathfrak{S}$ is a subalgebra of $\mathfrak{J}$, $\mathfrak{S} \cong \mathfrak{J}/\mathfrak{N}$.

The proof, in Penico [1], of this theorem is valid for characteristic $\neq 2$ since there are no central simple Jordan algebras for characteristic $p > 2$ different from those at characteristic 0. $\mathfrak{N}^2$ is not in general an ideal of $\mathfrak{J}$, but it turns out that $\mathfrak{J}\mathfrak{N}^2 + \mathfrak{N}^2$ is an ideal properly contained in $\mathfrak{N}$ (if $\mathfrak{N} \neq 0$), so the usual inductive argument may be used to reduce the proof to the case $\mathfrak{N}^2 = 0$. Here a lifting process by separate cases, similar to that in Lemmas 3.20 and 3.21, is employed. Some of Penico's computations may be eliminated by a method sketched in Taft [1]. However, it would be desirable to have a proof, if only for characteristic 0, which does not depend upon the classification of central simple Jordan algebras. The second Whitehead lemma for Jordan algebras of characteristic $\neq 2$ is valid, since it is equivalent to the case $\mathfrak{N}^2 = 0$ of the Wedderburn principal theorem. For its statement we require the notion of a Jordan bimodule, which is the analogue of an alternative bimodule.

Let $\mathfrak{J}$ be a Jordan algebra over F of characteristic $\neq 2$, and $\mathfrak{M}$ be a vector space over F. Then $\mathfrak{M}$ is a *Jordan bimodule* for $\mathfrak{J}$ in case there are two bilinear compositions ma, am in $\mathfrak{M}$ (m in $\mathfrak{M}$, a in $\mathfrak{J}$) satisfying

$$ma = am \qquad \text{for all} \quad a \text{ in } \mathfrak{J}, \quad m \text{ in } \mathfrak{M},$$

and

$$(4.36) \qquad (a^2, m, a) = (a^2, b, m) + 2(ma, b, a) = 0$$

for all a, b in $\mathfrak{J}$, m in $\mathfrak{M}$. Linearization of (4.36) gives

$$(4.37) \qquad \begin{aligned} (ab, m, c) + (bc, m, a) + (ca, m, b) &= 0, \\ (ac, b, m) + (ma, b, c) + (mc, b, a) &= 0 \end{aligned}$$

for all a, b, c in $\mathfrak{J}$, m in $\mathfrak{M}$. The vector space direct sum $\mathfrak{J} + \mathfrak{M}$ is made into a Jordan algebra (the *split null extension* or *semidirect sum*), by defining multiplication by

$$(a + m_1)(b + m_2) = ab + (m_1 b + am_2)$$

for all a, b in $\mathfrak{J}$, m_i in $\mathfrak{M}$; $\mathfrak{M}$ is an ideal of $\mathfrak{J} + \mathfrak{M}$, and $\mathfrak{M}^2 = 0$.

The linear operators S_a (a in $\mathfrak{J}$) defined by

$$S_a : m \to ma = am \qquad \text{for all} \quad m \text{ in } \mathfrak{M}$$

satisfy

(4.38) $\quad S_{a^2} S_a = S_a S_{a^2}, \qquad S_{a^2 b} - S_b S_{a^2} + 2S_a S_b S_a - 2S_a S_{ba} = 0$

for all a, b in $\mathfrak{J}$, and the mapping $a \to S_a$ is called a *representation* of $\mathfrak{J}$. Conversely, any linear mapping $a \to S_a$ of $\mathfrak{J}$ into some $\mathfrak{E}(\mathfrak{M})$ which satisfies (4.38) yields a Jordan bimodule $\mathfrak{M}$ when compositions in $\mathfrak{M}$ are defined by $ma = am = mS_a$ (a in $\mathfrak{J}$, m in $\mathfrak{M}$). The right multiplications R_a of $\mathfrak{J}$ give rise to the *regular representation* of $\mathfrak{J}$.

The *second Whitehead lemma for Jordan algebras* may be stated as follows: Let $\mathfrak{J}$ be a finite-dimensional separable Jordan algebra of characteristic $\neq 2$, and let $\mathfrak{M}$ be a Jordan bimodule for $\mathfrak{J}$. Let f be a bilinear mapping of $\mathfrak{J}$ into $\mathfrak{M}$ satisfying two conditions:

$$f(a, b) = f(b, a) \qquad \text{for all} \quad a, b \text{ in } \mathfrak{J}$$

and

$$F(a^2, b, a) = -(f(a, a), b, a) \qquad \text{for all} \quad a, b \text{ in } \mathfrak{J}$$

where

$$F(a, b, c) = f(a, b)c + f(ab, c) - af(b, c) - f(a, bc).$$

Then there exists a linear mapping g of $\mathfrak{J}$ into $\mathfrak{M}$ such that

$$f(a, b) = ag(b) + g(a)b - g(ab) \qquad \text{for all} \quad a, b \text{ in } \mathfrak{J}.$$

That the second Whitehead lemma is equivalent to the case $\mathfrak{N}^2 = 0$ of the Wedderburn principal theorem may be seen in exactly the same way that Proposition 3.22 was shown to be equivalent to the case $\mathfrak{N}^2 = 0$ of Theorem 3.18.

3. DERIVATIONS; SIMPLE LIE ALGEBRAS OF TYPE F

Since 1950 there has been considerable research activity on the relationships among Cayley algebras and Cayley planes, exceptional central simple Jordan algebras, and the exceptional simple Lie algebras and groups. We have listed a number of these papers in the bibliography, but we do not try to duplicate the admirable chronology and the references which appear in Freudenthal [4].

Let $\mathfrak{C}$ be any Cayley algebra over F, and let $\mathfrak{J}$ be the exceptional central simple Jordan algebra $\mathfrak{J} = \mathfrak{H}(\mathfrak{C}_3)$. Write x in (4.22) as

$$(4.39) \qquad x = \xi_1 e_1 + \xi_2 e_2 + \xi_3 e_3 + a_1 + b_2 + c_3$$

(ξ_i in F; a, b, c in $\mathfrak{C}$). Then we have the Peirce decomposition

$$(4.40) \qquad \mathfrak{J} = Fe_1 + Fe_2 + Fe_3 + \mathfrak{J}_1 + \mathfrak{J}_2 + \mathfrak{J}_3$$

of $\mathfrak{J} = \mathfrak{H}(\mathfrak{C}_3)$ where (4.20) implies

$$\mathfrak{J}_{ij} = \mathfrak{J}_k = \{d_k \,|\, d \text{ in } \mathfrak{C}\}, \qquad i, j, k \text{ distinct.}$$

The multiplicative rules (4.26)–(4.31) hold. Commutativity, bilinearity, and the multiplicative rules (4.26)–(4.31) characterize the multiplication in $\mathfrak{J} = \mathfrak{H}(\mathfrak{C}_3)$. We can without ambiguity drop the $\cdot$ in the products, writing xy for the product in $\mathfrak{J}$. As in the proof of Theorem 4.8, we denote by u the unity element in $\mathfrak{C}$, reserving the notation 1 for the unity element $1 = e_1 + e_2 + e_3$ in $\mathfrak{J}$.

Any element x in $\mathfrak{J} = \mathfrak{H}(\mathfrak{C}_3)$ satisfies a cubic equation

$$(4.41) \quad x^3 - T(x)x^2 + Q(x)x - N(x)1 = 0, \qquad T(x), Q(x), N(x) \text{ in } F,$$

where $x^2 x = x x^2 (= x^3)$ since $\mathfrak{J}$ is commutative, and x in (4.22) implies (4.41) where

$$(4.42) \qquad T(x) = \xi_1 + \xi_2 + \xi_3,$$

$$Q(x) = \xi_1 \xi_2 + \xi_2 \xi_3 + \xi_3 \xi_1 - n(a) - n(b) - n(c)$$

$$= \tfrac{1}{2}[(T(x))^2 - T(x^2)],$$

and

$$(4.43) \qquad N(x) = \xi_1 \xi_2 \xi_3 - \xi_1 n(a) - \xi_2 n(b) - \xi_3 n(c) + t(abc).$$

If F has characteristic $\neq 3$ (as well as $\neq 2$), formula (4.43) may be written as

$$N(x) = \tfrac{1}{6}[(T(x))^3 - 3T(x)T(x^2) + 2T(x^3)].$$

Taking a basis for $\mathfrak{J}$ adapted to the Peirce decomposition (4.40), we can compute the diagonal elements of a matrix for R_x, x in (4.39), easily from (4.26) to (4.31), obtaining

$$(4.44) \qquad\qquad \text{trace } R_x = 9T(x) \qquad \text{for all } x \text{ in } \mathfrak{J}.$$

Define

(4.45) $\langle x, y \rangle = T(xy)$ for all x, y in $\mathfrak{J}$.

For x in (4.39) and

$$y = \eta_1 e_1 + \eta_2 e_2 + \eta_3 e_3 + f_1 + g_2 + h_3 \qquad (\eta_i \text{ in } F; \ f, g, h \text{ in } \mathfrak{C})$$

we have

(4.46) $\langle x, y \rangle = \sum_{i=1}^{3} \xi_i \eta_i + 2(a, f) + 2(b, g) + 2(c, h),$

where (,) is the nondegenerate symmetric bilinear form (3.60) on $\mathfrak{C}$.
Clearly $\langle x, y \rangle$ is a symmetric bilinear form on $\mathfrak{J}$. Also $\langle x, y \rangle$ is non-
degenerate: $\langle x, y \rangle = 0$ for all y in $\mathfrak{J}$ implies $\xi_1 = \xi_2 = \xi_3 = (a, f) = (b, g) = (c, h) = 0$ $(i = 1, 2, 3)$ for all f, g, h in $\mathfrak{C}$. Hence $a = b = c = 0$
since (,) is nondegenerate on $\mathfrak{C}$. That is, $x = 0$. Also $\langle x, y \rangle$ is a trace
form on $\mathfrak{J}$, as we can see easily from (4.44) if F has characteristic $\neq 3$.
For

$$9\langle xy, z \rangle - 9\langle x, yz \rangle = 9T((x, y, z))$$

$$= \text{trace } R_{(x,y,z)} = \text{trace}[R_y, [R_x, R_z]] = 0$$

for all x, y, z in $\mathfrak{J}$ by (4.6). *Assume that F has characteristic $\neq 3$*. Then

(4.47) $\langle x, y, z \rangle = \langle xy, z \rangle$ $(= \langle x, yz \rangle)$ for all x, y, z in $\mathfrak{J}$

defines a symmetric trilinear form on $\mathfrak{J}$. A linear operator A on $\mathfrak{J}$ is
skew relative to (4.46) in case

(4.48) $\langle xA, y \rangle + \langle x, yA \rangle = 0$ for all x, y in $\mathfrak{J}$,

and we write $A \in \mathfrak{o}(27, q)$ where $q(x) = T(x^2)$. Also $\langle x, y \rangle$ is said to be
(*Lie*) *invariant* under A in case (4.48) holds. Note that (4.48) is equivalent
to $\langle xA, x \rangle = 0$ for all x in $\mathfrak{J}$. Similarly,

(4.49) $\langle xA, y, z \rangle + \langle x, yA, z \rangle + \langle x, y, zA \rangle = 0$ for all x, y, z in $\mathfrak{J}$

is equivalent to $\langle xA, x^2 \rangle$ $(= \langle xA, x, x \rangle) = 0$ for all x in $\mathfrak{J}$; $\langle x, y, z \rangle$
is (*Lie*) *invariant* under A in case (4.49) holds.

We now prove that a linear operator D on $\mathfrak{J}$ is a derivation of $\mathfrak{J}$ if and only if D leaves both $\langle x, y \rangle$ and $\langle x, y, z \rangle$ invariant. Let $\mathfrak{J}_0$ be the 26-dimensional subspace

$$\mathfrak{J}_0 = \{x \text{ in } \mathfrak{J} \mid T(x) = 0\}$$

of $\mathfrak{J}$. We have

(4.50) $\mathfrak{J}D \subseteq \mathfrak{J}_0$ for all D in $\mathfrak{D}(\mathfrak{J})$.

For trace $R_{xD} = \text{trace } [R_x, D] = 0$ for all x in $\mathfrak{J}$, D in $\mathfrak{D}(\mathfrak{J})$, implying $T(xD) = 0$ by (4.44). It follows that D is in $\mathfrak{o}(27, q)$ for all D in $\mathfrak{D}(\mathfrak{J})$:

$$\langle xD, y \rangle + \langle x, yD \rangle = T((xD)y + x(yD))$$

$$= T((xy)D) = 0$$

for all x, y in $\mathfrak{J}$. Also D in $\mathfrak{D}(\mathfrak{J})$ implies

$$\langle xD, y, z \rangle + \langle x, yD, z \rangle + \langle x, y, zD \rangle$$

$$= T((xD)(yz) + x((yD)z + y(zD))) = T((x(yz))D) = 0$$

for all x, y, z in $\mathfrak{J}$. Conversely, if D in $\mathfrak{E}(\mathfrak{J})$ satisfies (4.48) and (4.49), then $D^* = -D$ and

$$0 = \langle (xD)y, z \rangle + \langle x(yD), z \rangle + \langle xy, zD \rangle$$

$$= \langle (xD)y + x(yD) - (xy)D, z \rangle$$

for all x, y, z in $\mathfrak{J}$ by (4.47). Since $\langle \, , \, \rangle$ is nondegenerate on $\mathfrak{J}$, we have $(xD)y + x(yD) - (xy)D = 0$, or D is in $\mathfrak{D}(\mathfrak{J})$.

Over an algebraically closed field of characteristic 0 there is exactly one 52-dimensional simple Lie algebra, the exceptional algebra F_4 (Jacobson [25], pp. 135–146). We say that a (central simple) Lie algebra $\mathfrak{L}$ over F of characteristic 0 is of *type F* in case, for the algebraic closure K of F, we have $\mathfrak{L}_K \cong F_4$. In Theorem 4.11 below we see that $F_4 \cong \mathfrak{D}(\mathfrak{J})$ for $\mathfrak{J} = \mathfrak{H}(\mathfrak{C}_3)$. For F of characteristic $\neq 2, 3$ we extend the definition above, defining $\mathfrak{L}$ to be of *type F* in case, for the algebraic closure K of F, we have $\mathfrak{L}_K \cong \mathfrak{D}(\mathfrak{J})$ where $\mathfrak{J} = \mathfrak{H}(\mathfrak{C}_3)$.

Theorem 4.9 (Chevalley-Schafer). *Let $\mathfrak{J}$ be an exceptional central simple Jordan algebra over F of characteristic $\neq 2, 3$. Then the derivation algebra $\mathfrak{D}(\mathfrak{J})$ of $\mathfrak{J}$ is a 52-dimensional central simple Lie algebra (of type F).*

Proof. Since $\mathfrak{D}(\mathfrak{J})_K = \mathfrak{D}(\mathfrak{J}_K)$ for any extension K of F, it is sufficient to prove that $\mathfrak{D}(\mathfrak{J})$ is 52-dimensional and simple for $\mathfrak{J} = \mathfrak{H}(\mathbb{C}_3)$. Let

$$(4.51) \qquad \mathfrak{D}_0 = \{D \text{ in } \mathfrak{D}(\mathfrak{J}) \,|\, e_1 D = e_2 D \,(= e_3 D) = 0\}.$$

Then

$$(4.52) \qquad \mathfrak{J}_i D \subseteq \mathfrak{J}_i \qquad \text{for all} \quad D \text{ in } \mathfrak{D}_0 \qquad (i = 1, 2, 3),$$

since (4.28) implies that $0 = (e_i a_i)D = e_i(a_i D)$, or $a_i D$ is in $\mathfrak{J}_i + Fe_j + Fe_k$ for all a_i in $\mathfrak{J}_i$ and D in $\mathfrak{D}_0$ $(i, j, k$ distinct), while $\frac{1}{2}a_i D = (e_j a_i)D = e_j(a_i D)$ by (4.29), yielding $a_i D$ in $\mathfrak{J}_i + \mathfrak{J}_k$. This implies (4.52). For any D in $\mathfrak{D}_0$, let U_i be the linear operator on $\mathbb{C}$ defined by

$$(4.53) \qquad a_i D = (aU_i)_i \qquad \text{for all} \quad a \text{ in } \mathbb{C} \quad (a_i \text{ in } \mathfrak{J}_i).$$

Then (4.30) implies

$$(a_i b_i)D = (a, b)(e_j + e_k)D = 0 = (a_i D)b_i + a_i(b_i D)$$
$$= (aU_i)_i b_i + a_i(bU_i)_i = [(aU_i, b) + (a, bU_i)](e_j + e_k),$$

so that $(aU_i, b) + (a, bU_i) = 0$ for all a, b in $\mathbb{C}$; that is, U_i is in $\mathfrak{o}(8, n)$. The involution $S : a \to \bar{a} = aS$ of $\mathbb{C}$ satisfies $S^* = S = S^{-1}$. For

$$(aS, b) = (\bar{a}, b) = \tfrac{1}{2}t(\bar{a}b) = \tfrac{1}{2}t(\bar{b}\bar{\bar{a}}) = (\bar{b}, a) = (a, bS)$$

for all a, b in $\mathbb{C}$. Then

$$(S^{-1}U_i S)^* = S^* U_i^* S^{*-1} = -S^{-1}U_i S.$$

Hence $S^{-1}U_i S$ is in $\mathfrak{o}(8, n)$ for U_i in (4.53) $(i = 1, 2, 3)$. Let $U = S^{-1}U_1 S \in \mathfrak{o}(8, n)$. Then U_2 and U_3 in (4.53) are the unique $U_2 = U'$, $U_3 = U''$ given by the principle of triality (Theorem 3.31). For (4.31) and (4.53) imply

$$2(a_i b_j)D = 2(a_i D)b_j + 2a_i(b_j D) = 2(aU_i)_i b_j + 2a_i(bU_j)_j$$
$$= (\bar{b}(aU_i S))_k + ((bU_j S)\bar{a})_k = (\bar{b}\bar{a})_k D = ((\bar{b}\bar{a})U_k)_k$$

for any cyclic permutation i, j, k of $1, 2, 3$. That is, $\bar{b}(aU_i S) + (bU_j S)\bar{a} = (\bar{b}\bar{a})U_k$, or

$$(ab)S^{-1}U_k S = (aU_i)b + a(bU_j) \qquad \text{for all} \quad a, b \text{ in } \mathfrak{C},$$

for any cyclic permutation i, j, k of $1, 2, 3$. In particular, we have

$$(ab)U = (ab)S^{-1}U_1 S = (aU_2)b + a(bU_3),$$

implying $U_2 = U'$, $U_3 = U''$ by the principle of triality. Conversely, if U is any element of $\mathfrak{o}(8, n)$, the linear operator D_U on $\mathfrak{J}$ defined by $e_i D_U = 0$ $(i = 1, 2, 3)$ and

$$(4.54) \qquad U_1 = S^{-1}US, \qquad U_2 = U', \qquad U_3 = U''$$

in (4.53) is in $\mathfrak{D}_0$. For we need only check that

$$\langle xD_U, x \rangle = 0 \qquad \text{for all} \quad x \text{ in } \mathfrak{J}$$

and

$$\langle xD_U, x^2 \rangle = 0 \qquad \text{for all} \quad x \text{ in } \mathfrak{J},$$

where

$$xD_U = (aU_1)_1 + (bU_2)_2 + (cU_3)_3, \qquad x \text{ in } (4.39),$$

with U_i given by (4.54). Now (4.46) implies that

$$\langle xD_U, x \rangle = 2(aU_1, a) + 2(bU_2, b) + 2(cU_3, c) = 0$$

since each U_i is in $\mathfrak{o}(8, n)$. The component of x^2 in $\mathfrak{J}_1 + \mathfrak{J}_2 + \mathfrak{J}_3$ is

$$(\xi_2 + \xi_3)a_1 + (\bar{c}\bar{b})_1 + (\xi_1 + \xi_3)b_2 + (\bar{a}\bar{c})_2 + (\xi_1 + \xi_2)c_3 + (\bar{b}\bar{a})_3,$$

so

$$\begin{aligned}
\langle xD_U, x^2 \rangle &= (\xi_2 + \xi_3)(aU_1, a) + 2(aU_1, \bar{c}\bar{b}) + (\xi_1 + \xi_3)(bU_2, b) \\
&\quad + 2(bU_2, \bar{a}\bar{c}) + (\xi_1 + \xi_2)(cU_3, c) + 2(cU_3, \bar{b}\bar{a}) \\
&= 2(\bar{a}US, (bc)S) + 2(bU', \overline{ca}) + 2(cU'', \overline{ab}) \\
&= 2(\bar{a}U, bc) + t((bU')ca) + t((cU'')ab) \\
&= -2(\bar{a}, (bc)U) + t((bU')ca) + t(b(cU'')a) \\
&= -2(\bar{a}, (bU')c) - 2(\bar{a}, b(cU'')) \\
&\quad + 2((bU')c, \bar{a}) + 2(b(cU''), \bar{a}) = 0.
\end{aligned}$$

Hence D_U is a derivation of $\mathfrak{J}$, and D_U is in $\mathfrak{D}_0$. Now $U \to D_U$ is an isomorphism of $\mathfrak{o}(8, n)$ onto $\mathfrak{D}_0$. For it remains only to show that

$$[D_T, D_U] = D_{[T,U]} \qquad \text{for all} \quad T, U \text{ in } \mathfrak{o}(8, n).$$

For a_i in $\mathfrak{J}_i$, we have $a_i[D_T, D_U] = (a[T_i, U_i])_i = (a[T, U]_i)_i = a_i D_{[T,U]}$, since

(4.55) $[T_i, U_i] = [T, U]_i \qquad (i = 1, 2, 3) \qquad \text{for all} \quad T, U \text{ in } \mathfrak{o}(8, n).$

The case $i = 1$ of (4.55) is immediate from the first part of (4.54), while $(ab)[T, U] = (a[T', U'])b + a(b[T'', U''])$ for all a, b in $\mathfrak{C}$ implies $[T, U]' = [T', U']$, $[T, U]'' = [T'', U'']$ by the principle of triality.

It is well known (and easy to prove) that the 28-dimensional (orthogonal) Lie algebra $\mathfrak{o}(8, n)$ is simple. Hence $\mathfrak{D}_0$ is simple. Also, since the associative enveloping algebra of $\mathfrak{o}(8, n)$ is $\mathfrak{E}(\mathfrak{C})$, we have

(4.56) $\mathfrak{C}\,\mathfrak{o}(8, n) = \mathfrak{C}.$

We use these facts to show that $\mathfrak{D}(\mathfrak{J})$ is simple. Let

$$\mathfrak{D}_i = \{[R_{a_i}, R_{e_j - e_k}] \mid a_i \in \mathfrak{J}_i\}$$

(i, j, k cyclic permutation of 1, 2, 3). Note that

(4.57)
$$e_i[R_{a_i}, R_{e_j - e_k}] = 0, \qquad e_j[R_{a_i}, R_{e_j - e_k}] = -\tfrac{1}{2}a_i,$$
$$e_k[R_{a_i}, R_{e_j - e_k}] = \tfrac{1}{2}a_i \qquad (i, j, k \text{ distinct})$$

by (4.26)–(4.29). Then dim $\mathfrak{D}_i = 8$, so that $\mathfrak{D}_0 + \mathfrak{D}_1 + \mathfrak{D}_2 + \mathfrak{D}_3$ is a subspace of $\mathfrak{D}(\mathfrak{J})$ of dimension $\leq 52 = 28 + 8 + 8 + 8$. Then $\mathfrak{D}_0 + \mathfrak{D}_1 + \mathfrak{D}_2 + \mathfrak{D}_3$ is a (52-dimensional) vector space direct sum. For any element D in this space may be written in the form

(4.58) $D = \tilde{D} + [R_{a_1}, R_{e_2 - e_3}] + [R_{b_2}, R_{e_3 - e_1}] + [R_{c_3}, R_{e_1 - e_2}]$

with $\tilde{D}$ in $\mathfrak{D}_0$; a, b, c in $\mathfrak{C}$. If D in (4.58) is 0, then $0 = e_1 D = \tfrac{1}{2}(b_2 - c_3)$ by (4.57), implying $b_2 = c_3 \in \mathfrak{J}_2 \cap \mathfrak{J}_3 = 0$, so that $b = c = 0$. Similarly, $0 = e_2 D = \tfrac{1}{2}(c_3 - a_1)$ implies $a = 0$. Hence $\tilde{D} = 0$. Finally,

(4.59) $\mathfrak{D}(\mathfrak{J}) = \mathfrak{D}_0 + \mathfrak{D}_1 + \mathfrak{D}_2 + \mathfrak{D}_3$ (direct sum).

For, if D is any element of $\mathfrak{D}(\mathfrak{J})$, then $e_i D = (e_i^2)D = 2e_i(e_i D)$ implies $e_i D$ is in $\mathfrak{J}_j + \mathfrak{J}_k$ (i, j, k distinct). Write

$$e_1 D = \tfrac{1}{2}b_2 + \tfrac{1}{2}f_3, \qquad e_2 D = \tfrac{1}{2}c_3 + \tfrac{1}{2}g_1, \qquad e_3 D = \tfrac{1}{2}a_1 + \tfrac{1}{2}h_2$$

for a, b, c, f, g, h in $\mathfrak{C}$. Then

$$0 = (e_1 e_2)D = (e_1 D)e_2 + e_1(e_2 D) = \tfrac{1}{4}f_3 + \tfrac{1}{4}c_3,$$

so that $f = -c$. Similarly, $g = -a$. Using $1D = 0$, we have $h = -b$, implying

$$(4.60) \quad e_1 D = \tfrac{1}{2}(b_2 - c_3), \qquad e_2 D = \tfrac{1}{2}(c_3 - a_1), \qquad e_3 D = \tfrac{1}{2}(a_1 - b_2)$$

for a, b, c in $\mathfrak{C}$. For D in $\mathfrak{D}(\mathfrak{J})$ define $\tilde{D}$ by (4.58) where a, b, c in $\mathfrak{C}$ are given by (4.60). We have $e_1 \tilde{D} = e_1 D - \tfrac{1}{2}(b_2 - c_3) = 0$ by the same computation as before. Similarly, $e_2 \tilde{D} = 0$. Hence $\tilde{D}$ is in $\mathfrak{D}_0$, and D is in $\mathfrak{D}_0 + \mathfrak{D}_1 + \mathfrak{D}_2 + \mathfrak{D}_3$, implying (4.59). Now

$$(4.61) \qquad [\mathfrak{D}_i, \mathfrak{D}_0] = \mathfrak{D}_i \qquad (i = 1, 2, 3).$$

For

$$(4.62) \qquad [[R_x, R_y], D] = [R_{xD}, R_y] + [R_x, R_{yD}]$$

for all D in $\mathfrak{D}(\mathfrak{J})$ and x, y in $\mathfrak{J}$ by the Jacobi identity. Hence D in $\mathfrak{D}_0$ and a_i in $\mathfrak{J}_i$ imply

$$[[R_{a_i}, R_{e_j - e_k}], D] = [R_{a_i D}, R_{e_j - e_k}] + [R_{a_i}, R_{(e_j - e_k)D}]$$
$$= [R_{(aU_i)_i}, R_{e_j - e_k}]$$

is in $\mathfrak{D}_i$, so that $[\mathfrak{D}_i, \mathfrak{D}_0] \subseteq \mathfrak{D}_i$. Then (4.56) implies (4.61) since U_i ranges over $\mathfrak{o}(8, n)$ as D ranges over $\mathfrak{D}_0$.

Also

$$(4.63) \qquad [\mathfrak{D}_i, \mathfrak{D}_i] \subseteq \mathfrak{D}_0 \qquad (i = 1, 2, 3)$$

and

$$(4.64) \qquad [\mathfrak{D}_i, \mathfrak{D}_j] \quad (= [\mathfrak{D}_j, \mathfrak{D}_i]) \subseteq \mathfrak{D}_k \qquad (i, j, k \text{ distinct}).$$

To prove (4.63) and (4.64), we first observe that

$$(4.65) \qquad \mathfrak{J}_k \mathfrak{D}_i \subseteq \mathfrak{J}_j, \qquad \mathfrak{J}_j \mathfrak{D}_i \subseteq \mathfrak{J}_k$$

for any cyclic permutation i, j, k of $1, 2, 3$. For

$$b_k[R_{a_i}, R_{e_j - e_k}] = (a_i, b_k, e_j - e_k) = -(a_i b_k)e_k - \tfrac{1}{2}a_i b_k = -a_i b_k \in \mathfrak{J}_j$$

by (4.31), (4.28), and (4.29), while similarly $b_j[R_{a_i}, R_{e_j - e_k}] \in \mathfrak{J}_k$. Also

$$(a_i, b_i, e_j - e_k) = (a, b)(e_j + e_k)(e_j - e_k) = (a, b)(e_j - e_k)$$

by (4.30) and (4.29), implying

$$b_i[R_{a_i}, R_{e_j - e_k}] = (a, b)(e_j - e_k),$$

(4.66)

$$\mathfrak{J}_i \mathfrak{D}_i \subseteq F e_j + F e_k.$$

Since (a, b) in (4.66) is nondegenerate, we see that

(4.67) $D_i \in \mathfrak{D}_i$ and $\mathfrak{J}_i D_i = 0$ imply $D_i = 0$.

To prove (4.63), we suppose that $D = \tilde{D} + D_i + D_j + D_k$ ($\tilde{D} \in \mathfrak{D}_0$, $D_h \in \mathfrak{D}_h$) is in $[\mathfrak{D}_i, \mathfrak{D}_i]$. Then (4.65) implies that $\mathfrak{J}_j D \subseteq \mathfrak{J}_j[\mathfrak{D}_i, \mathfrak{D}_i] \subseteq \mathfrak{J}_j$, so that $\mathfrak{J}_j D_j = 0$ by (4.52), (4.65), and (4.66). Then (4.67) implies that $D_j = 0$. Similarly, $D_k = 0$. Also

$$\mathfrak{J}_i D \subseteq \mathfrak{J}_i[\mathfrak{D}_i, \mathfrak{D}_i] \subseteq (\mathfrak{J}_i \mathfrak{D}_i)\mathfrak{D}_i \subseteq (F e_j + F e_k)\mathfrak{D}_i \subseteq \mathfrak{J}_i$$

by (4.65), (4.66), and (4.57). Hence $\mathfrak{J}_i D_i = 0$ by (4.52), (4.65), and (4.66), implying $D_i = 0$ by (4.67). Thus $D = \tilde{D} \in \mathfrak{D}_0$, implying (4.63). To prove (4.64), we suppose that $D = \tilde{D} + D_i + D_j + D_k$ ($\tilde{D} \in \mathfrak{D}_0$, $D_h \in \mathfrak{D}_h$) is in $[\mathfrak{D}_i, \mathfrak{D}_j]$. Now

$$\mathfrak{J}_i D \subseteq \mathfrak{J}_i[\mathfrak{D}_i, \mathfrak{D}_j] \subseteq (F e_j + F e_k)\mathfrak{D}_j + \mathfrak{J}_k \mathfrak{D}_i \subseteq \mathfrak{J}_j$$

in the same manner as above, implying $\mathfrak{J}_i \tilde{D} = \mathfrak{J}_i D_i = 0$. Similarly, $\mathfrak{J}_j \tilde{D} = \mathfrak{J}_j D_j = 0$. Hence (4.67) implies $D_i = D_j = 0$. Also $\mathfrak{J}_k D \subseteq \mathfrak{J}_k[\mathfrak{D}_i, \mathfrak{D}_j] \subseteq F e_1 + F e_2 + F e_3$, so that we have $\mathfrak{J}_k \tilde{D} = 0$ in the same manner. Hence $\mathfrak{J} \tilde{D} = (\mathfrak{J}_1 + \mathfrak{J}_2 + \mathfrak{J}_3)\tilde{D} = 0$, implying $\tilde{D} = 0$ and $D = D_k \in \mathfrak{D}_k$. This establishes (4.64).

Let $\mathfrak{B}$ be an ideal of $\mathfrak{D}(\mathfrak{J})$. We apply Lemma 3.27 with $\mathfrak{A} = \mathfrak{D}(\mathfrak{J})$, $\mathfrak{S} = \mathfrak{D}_0$, and $\mathfrak{T} = \mathfrak{D}_1 + \mathfrak{D}_2 + \mathfrak{D}_3$. We have dim $\mathfrak{D}_0 = 28 > 24 =$ dim $(\mathfrak{D}_1 + \mathfrak{D}_2 + \mathfrak{D}_3)$, and (iii) holds by (4.61). Lemma 3.27 implies that, if $\mathfrak{B} \neq \mathfrak{D}(\mathfrak{J})$, then

$$\mathfrak{B} \subseteq \mathfrak{D}_1 + \mathfrak{D}_2 + \mathfrak{D}_3.$$

Let $D = D_1 + D_2 + D_3$, D_i in $\mathfrak{D}_i$, be any element of $\mathfrak{B}$. If $A_i = [R_{a_i}, R_{e_j - e_k}]$ is any element of $\mathfrak{D}_i$ ($i = 1, 2, 3$), then $\mathfrak{B}$ contains $[D, A_i] = [D_i, A_i] + [D_j, A_i] + [D_k, A_i]$ in $[D_i, A_i] + \mathfrak{D}_k + \mathfrak{D}_j$ by (4.64). Hence $[D_i, A_i]$ is in $\mathfrak{D}_0 \cap (\mathfrak{D}_i + \mathfrak{D}_j + \mathfrak{D}_k) = 0$ by (4.63). For a fixed i, let $D_i = [R_{b_i}, R_{e_j - e_k}]$ for some b in $\mathfrak{C}$. For all c in $\mathfrak{C}$, (4.66) implies that $c_i A_i = (a, c)(e_j - e_k)$ and $c_i D_i = (b, c)(e_j - e_k)$, so that

$$0 = c_i[D_i, A_i] = (b, c)(e_j - e_k)A_i - (a, c)(e_j - e_k)D_i$$
$$= -(b, c)a_i + (a, c)b_i$$

by (4.57). That is,

(4.68) $(a, c)b = (b, c)a$ for all a, c in $\mathfrak{C}$.

Choose a in $\mathfrak{C}$ independent of b. Then (4.68) implies that $(b, c) = 0$ for all c in $\mathfrak{C}$. Hence $b = 0$, $b_i = 0$ and $D_i = 0$. Since $D_i = 0$ for $i = 1, 2, 3$, we have $D = 0$, implying $\mathfrak{B} = 0$. That is, $\mathfrak{D}(\mathfrak{J})$ is simple. This completes the proof of Theorem 4.9.

Corollary 4.10. *Every derivation D of an exceptional central simple algebra $\mathfrak{J}$ of characteristic $\neq 2, 3$ is inner:*

$$D = \sum [R_{x_i}, R_{z_i}]; \qquad x_i, z_i \text{ in } \mathfrak{J}.$$

Proof. By (4.62) the inner derivations $\sum [R_{x_i}, R_{z_i}]$ form an ideal in the simple algebra $\mathfrak{D}(\mathfrak{J})$. This ideal is $\neq 0$ by (4.57).

We omit the proof of the fact that every central simple Lie algebra of type F (over a field of characteristic $\neq 2, 3$) is isomorphic to $\mathfrak{D}(\mathfrak{J})$ for some exceptional central simple Jordan algebra $\mathfrak{J}$, and that $\mathfrak{D}(\mathfrak{J}) \cong \mathfrak{D}(\tilde{\mathfrak{J}})$ if and only if $\mathfrak{J} \cong \tilde{\mathfrak{J}}$ (Tomber [1]; Barnes [1]; Seligman [3]).

Jacobson's Theorem 2.5 implies the analogue of Corollary 3.30 for Jordan algebras of characteristic 0: Every derivation D of a finite-dimensional semisimple Jordan algebra $\mathfrak{J}$ of characteristic 0 is inner:

$$D = \sum [R_{x_i}, R_{z_i}]; \qquad x_i, z_i \text{ in } \mathfrak{J}.$$

In order to extend this to separable algebras over fields of characteristic $\neq 2$, one needs to add the hypothesis that the characteristic of F does not divide the degree over its center of any special simple summand of $\mathfrak{J}$. The generalization embodied in the *first Whitehead lemma for Jordan algebras* may be stated as follows (Harris [2]). Let $\mathfrak{J}$ be a finite-dimensional separable Jordan algebra over F of characteristic $\neq 2$, such that the characteristic of F does not divide the degree over its center of any special simple summand of $\mathfrak{J}$. Let $\mathfrak{M}$ be a Jordan bimodule for $\mathfrak{J}$, and let $\mathfrak{B} = \mathfrak{J} + \mathfrak{M}$ be the split null extension. Let f be a one-cocycle of $\mathfrak{J}$ into $\mathfrak{M}$; that is, a linear transformation of $\mathfrak{J}$ into $\mathfrak{M}$ satisfying

$$f(ab) = f(a)b + af(b) \qquad \text{for all} \quad a, b \text{ in } \mathfrak{J}.$$

Then there exist x_i in $\mathfrak{J}$, z_i in $\mathfrak{M}$ satisfying

$$f(a) = \sum(x_i, a, z_i) \qquad \text{for all} \quad a \text{ in } \mathfrak{J}.$$

The first Whitehead lemma is equivalent to the statement that, if $\mathfrak{J}$ is a separable subalgebra of a finite-dimensional Jordan algebra $\mathfrak{B}$ where the characteristic satisfies the conditions above, then any derivation of $\mathfrak{J}$ into $\mathfrak{B}$ may be extended to an inner derivation $D = \sum[R_{x_i}, R_{z_i}]$ of $\mathfrak{B}$.

Just as for alternative algebras, the first Whitehead lemma for Jordan algebras may be used to prove the following analogue of the Malcev-Harish-Chandra theorem for Lie algebras. Let $\mathfrak{J}$ be a finite-dimensional Jordan algebra of characteristic 0 with Wedderburn decomposition $\mathfrak{J} = \mathfrak{S} + \mathfrak{N}$, and let $\mathfrak{B}$ be a semisimple subalgebra of $\mathfrak{J}$. Then there exists a (nilpotent) derivation D of $\mathfrak{J}$ in the radical of the multiplication algebra $\mathfrak{M}(\mathfrak{J})$ such that the automorphism $G = \exp D$ of $\mathfrak{J}$ carries $\mathfrak{B}$ onto a subalgebra of $\mathfrak{S}$. We have the corollary: If $\mathfrak{J}$ has Wedderburn decompositions $\mathfrak{J} = \mathfrak{S} + \mathfrak{N} = \mathfrak{S}_1 + \mathfrak{N}$, then $\mathfrak{S}$ and $\mathfrak{S}_1$ are conjugate under an automorphism $G = \exp D$ of the type above (Jacobson [9]).

4. SIMPLE LIE ALGEBRAS OF TYPE E_6

Over an algebraically closed field K of characteristic 0 there are five exceptional simple Lie algebras: the 14-dimensional algebra G_2 and the 52-dimensional algebra F_4 already discussed, and three others (E_6, E_7, E_8) of dimensions 78, 133, and 248, respectively. In Theorem 4.12 we represent E_6 in terms of $\mathfrak{J} = \mathfrak{H}(\mathfrak{C}_3)$ as follows: $E_6 = \mathfrak{D}(\mathfrak{J}) + R_0(\mathfrak{J})$ where $R_0(\mathfrak{J}) = \{R_x \mid x \text{ in } \mathfrak{J}_0\}$. However, E_6 is not the only 78-dimensional simple Lie algebra over K; there are also the orthogonal Lie algebra $\mathfrak{o}(13)$ and the symplectic Lie algebra $\mathfrak{sp}(12)$ (denoted by B_6 and C_6, respectively, in Jacobson [25], p. 141). To distinguish E_6 from $\mathfrak{o}(13)$ and $\mathfrak{sp}(12)$ we use a fact, the proof of which lies outside of the scope of this book: these two latter 78-dimensional simple Lie algebras over K do not have irreducible representations of degree 27. [This follows from Chapters VII and VIII of Jacobson [25].] However, as we prove in

Lemma 4.11 below, the representation in (the 27-dimensional Jordan algebra) $\mathfrak{J}$ of the 78-dimensional simple Lie algebra $\mathfrak{D}(\mathfrak{J}) + R_0(\mathfrak{J})$ over K is irreducible. Hence this Lie algebra over K is E_6. A Lie algebra $\mathfrak{L}$ over F of characteristic 0 is said to be of *type* E_6 in case $\mathfrak{L}_K \cong E_6$ where K is the algebraic closure of F. Extending this definition to fields F of characteristic $\neq 2$, 3, we say that $\mathfrak{L}$ is of *type* E_6 over F in case $\mathfrak{L}_K \cong \mathfrak{D}(\mathfrak{J}) + R_0(\mathfrak{J})$ where $\mathfrak{J} = \mathfrak{H}(\mathfrak{C}_3)$ over the algebraic closure K of F.

Lemma 4.11. *Let* $\mathfrak{C}$ *be a Cayley algebra over* F *of characteristic* $\neq 2$, 3 *and* $\mathfrak{J} = \mathfrak{H}(\mathfrak{C}_3)$. *Then* $\mathfrak{D}(\mathfrak{J}) + R_0(\mathfrak{J})$ *acts irreducibly in* $\mathfrak{J}$.

Proof. By the remarks following (4.6), we know that the Lie multiplication algebra $\mathfrak{L}(\mathfrak{J})$ is $\mathfrak{L}(\mathfrak{J}) = R(\mathfrak{J}) + [R(\mathfrak{J}), R(\mathfrak{J})]$. Also $[R_0(\mathfrak{J}), \mathfrak{D}(\mathfrak{J})] \subseteq R_0(\mathfrak{J})$, since $[R_x, D] = R_{xD}$ and $T(xD) = \frac{1}{9}$ trace $R_{xD} = 0$. Hence the Lie enveloping algebra of $R_0(\mathfrak{J})$ is $\mathfrak{D}(\mathfrak{J}) + R_0(\mathfrak{J})$. For any element x of $\mathfrak{J}$ may be written as $x = \alpha 1 + x'$, α in F, x' in $\mathfrak{J}_0$. Then Corollary 4.10 implies that D in $\mathfrak{D}(\mathfrak{J})$ may be written as

$$D = \sum [R_{x_i}, R_{z_i}] = \sum [R_{x_{i'}}, R_{z_{i'}}] \quad \text{in} \quad [R_0(\mathfrak{J}), R_0(\mathfrak{J})].$$

Hence

$$(4.69) \qquad\qquad \mathfrak{D}(\mathfrak{J}) = [R_0(\mathfrak{J}), R_0(\mathfrak{J})].$$

Since $\mathfrak{J}$ is (right) simple, $\mathfrak{J}$ is irreducible relative to $R(\mathfrak{J})$. Hence $\mathfrak{J}$ is irreducible relative to $R_0(\mathfrak{J})$, and also relative to its Lie enveloping algebra $\mathfrak{D}(\mathfrak{J}) + R_0(\mathfrak{J})$.

Theorem 4.12 (Chevalley-Schafer). *Let* $\mathfrak{J}$ *be any finite-dimensional exceptional central simple Jordan algebra over* F *of characteristic* $\neq 2$, 3, *and* $\mathfrak{L} = \mathfrak{L}(\mathfrak{J})$ *be the Lie multiplication algebra of* $\mathfrak{J}$. *Then*

$$(4.70) \qquad\qquad \mathfrak{L} = F1_3 \oplus \mathfrak{L}'$$

where the derived algebra

$$\mathfrak{L}' = \mathfrak{D}(\mathfrak{J}) + \{R_x \mid \text{trace } R_x = 0\} \qquad \text{(direct sum)}$$

is a 78-dimensional central simple Lie algebra (of type E_6) *over* F. *If* $\mathfrak{J} = \mathfrak{H}(\mathfrak{C}_3)$, *then*

$$(4.71) \qquad\qquad \mathfrak{L}' = \mathfrak{D}(\mathfrak{J}) + R_0(\mathfrak{J}) \qquad \text{(direct sum)}.$$

Proof. We have $\mathfrak{L} = R(\mathfrak{J}) + [R(\mathfrak{J}), R(\mathfrak{J})]$. Also trace $1_\mathfrak{J} = 27 \neq 0$ implies $1_\mathfrak{J} \notin \mathfrak{L}'$; hence $F1_\mathfrak{J} \oplus \mathfrak{L}' \subseteq \mathfrak{L}$. For the remainder of the proof we may pass to a scalar extension, and may assume that $\mathfrak{J} = \mathfrak{H}(\mathbb{C}_3)$. Then $R_0(\mathfrak{J}) = \{R_x \,|\, \text{trace } R_x = 0\}$ by (4.44). Also $\mathfrak{J}_0$ is the *associator subspace* $\mathfrak{P}$ of $\mathfrak{J}$ spanned by all associators (x, y, z); x, y, z in $\mathfrak{J}$. For (4.57) implies that $\mathfrak{P} \supseteq \mathfrak{J}_1 + \mathfrak{J}_2 + \mathfrak{J}_3$. Also, if u is the unity element of $\mathbb{C}$, then (4.30) and (4.29) imply that $\mathfrak{P}$ contains

$$2(u_i, u_i, e_j) = 2e_j - (e_j + e_k) = e_j - e_k \qquad (i, j, k \text{ distinct}).$$

Hence $\mathfrak{P} = \mathfrak{J}_0$, and

(4.72) $$[R_0(\mathfrak{J}), [R_0(\mathfrak{J}), R_0(\mathfrak{J})]] = R_0(\mathfrak{J})$$

by (4.6). Hence

$$\mathfrak{L}' = [R_0(\mathfrak{J}), R_0(\mathfrak{J})] + [R_0(\mathfrak{J}), [R_0(\mathfrak{J}), R_0(\mathfrak{J})]] = \mathfrak{D}(\mathfrak{J}) + R_0(\mathfrak{J}).$$

Applying any element in $\mathfrak{L}'$ to 1 in $\mathfrak{J}$, we see that (4.71) is a vector space direct sum (of dimension $78 = 52 + 26$). Hence (4.70) holds. It remains only to show that $\mathfrak{D}(\mathfrak{J}) + R_0(\mathfrak{J})$ is simple.

We apply Lemma 3.27 with $\mathfrak{S} = \mathfrak{D}(\mathfrak{J})$, $\mathfrak{T} = R_0(\mathfrak{J})$. We have shown (i) above; Theorem 4.9 implies (ii), and (iii) is given by (4.69) and (4.72). It follows that, if $\mathfrak{B}$ is any proper ideal of (4.71), then $\mathfrak{B} \subseteq R_0(\mathfrak{J})$. Let R_c (c in $\mathfrak{J}_0$) be any element of $\mathfrak{B}$. Then $\mathfrak{B}$ contains $[R_c, R_w]$ for any w in $\mathfrak{J}_0$. That is, for any $z = \alpha 1 + w$ in $\mathfrak{J}$, we have $[R_c, R_z] = [R_c, R_w]$ in $\mathfrak{D}(\mathfrak{J}) \cap \mathfrak{B} \subseteq \mathfrak{D}(\mathfrak{J}) \cap R_0(\mathfrak{J}) = 0$. Hence $(c, y, z) = 0$ for all y, z in $\mathfrak{J}$. That is, c is in the *left nucleus* of $\mathfrak{J}$: $L_{cy} = L_y L_c$ for all y in $\mathfrak{J}$. (Equivalently, $L_c R_z = R_z L_c$ for all z in $\mathfrak{J}$.) In a commutative algebra this is sufficient to guarantee that c is in the center of $\mathfrak{J}$:

$$R_c = L_c, \qquad R_c R_y = L_c R_y = R_y L_c = R_y R_c = L_y L_c = L_{cy} = R_{cy}$$

for all y in $\mathfrak{J}$, yielding (2.7). But the center of $\mathfrak{J}$ is $F1$. Hence c is in $F1 \cap \mathfrak{J}_0 = 0$, $R_c = 0$, $\mathfrak{B} = 0$. That is, $\mathfrak{L}'$ in (4.71) is simple.

We conclude this chapter with a remarkable theorem of Tits which combines alternative and Jordan algebras in a very illuminating characterization of the exceptional simple Lie algebras. To state the theorem in full generality one needs the concept of the "generic trace" (Jacobson [16, I]) of a Jordan algebra $\mathfrak{J}$ of degree 3. However, by limiting

ourselves to central simple Jordan algebras, we may gloss over this by
using, as in the particular case (4.44), the trace

$$T(x) = \frac{3}{\dim \mathfrak{J}} \text{ trace } R_x \qquad \text{for all} \quad x \text{ in } \mathfrak{J}$$

for the central simple Jordan algebras $\mathfrak{J}$ of degree 3 and dimension 6
(type B), dimension 9 (type A), dimension 15 (type C) and dimension 27
(type E). Division by 5 for type C is not necessary when using the actual
generic trace.

Let F be any field of characteristic $\neq 2, 3$. Let $\mathfrak{A}$ be a composition
algebra over F, and denote by u the identity element of $\mathfrak{A}$. We know that

$$ab = (a, \bar{b})u + a * b \qquad \text{for all} \quad a, b \text{ in } \mathfrak{A},$$

where $(a, b) = \frac{1}{2}t(a\bar{b})$ in F is defined by (3.60), and $a * b = ab - (a, \bar{b})u$
is in $\mathfrak{A}_0 = (Fu)^{\perp} = \{a \text{ in } \mathfrak{A} \mid t(a) = 0\}$. Hence

(4.73) $$ab = -(a, b)u + a * b \qquad \text{for all} \quad a, b \text{ in } \mathfrak{A}_0.$$

Similarly, if $\mathfrak{J}$ is a central simple Jordan algebra of degree 3 over F
with identity element e, then

(4.74) $$xy = \frac{1}{3}\langle x, y \rangle e + x * y \qquad \text{for all} \quad x, y \text{ in } \mathfrak{J}$$

where $\langle x, y \rangle = T(xy)$ in F is defined in terms of the generic trace $T(x)$
on $\mathfrak{J}$, and $x * y = xy - \frac{1}{3}\langle x, y \rangle e$ is in $\mathfrak{J}_0 = (Fe)^{\perp} = \{x \text{ in } \mathfrak{J} \mid T(x) = 0\}$
since $T(e) = 3$. In each of these Jordan algebras, every x satisfies the
cubic equation (4.41), and $\langle x, y \rangle$ is a nondegenerate trace form. The
one-dimensional algebra $\mathfrak{J} = Fe$ may also be included in (4.74) by
taking $\mathfrak{J}_0 = 0$.

Let

(4.75) $$\mathfrak{L} = \mathfrak{D}(\mathfrak{A}) + \mathfrak{A}_0 \otimes \mathfrak{J}_0 + \mathfrak{D}(\mathfrak{J})$$

be the vector space direct sum of the derivation algebras $\mathfrak{D}(\mathfrak{A})$ and
$\mathfrak{D}(\mathfrak{J})$ and the tensor product $\mathfrak{A}_0 \otimes_F \mathfrak{J}_0$. Then $\mathfrak{L}$ is made into an algebra
over F by defining a multiplication $[,]$ in $\mathfrak{L}$ which is bilinear and anti-
commutative, which agrees with the ordinary commutator in $\mathfrak{D}(\mathfrak{A})$ and
$\mathfrak{D}(\mathfrak{J})$, and which satisfies

(4.76) $$[\mathfrak{D}(\mathfrak{A}), \mathfrak{D}(\mathfrak{J})] = 0,$$

(4.77) $[a \otimes x, D] = aD \otimes x$

for all D in $\mathfrak{D}(\mathfrak{A})$, a in $\mathfrak{A}_0$, x in $\mathfrak{J}_0$,

(4.78) $[a \otimes x, E] = a \otimes xE$

for all E in $\mathfrak{D}(\mathfrak{J})$, a in $\mathfrak{A}_0$, x in $\mathfrak{J}_0$,

and

(4.79) $[a \otimes x, b \otimes y] = \frac{1}{12}\langle x, y \rangle D_{a,b} + (a * b) \otimes (x * y)$
$$- (a, b)[R_x, R_y]$$

for all a, b in $\mathfrak{A}_0$; x, y in $\mathfrak{J}_0$; where $D_{a,b}$ in $\mathfrak{D}(\mathfrak{A})$ is defined by (3.70). We have used

$$D_{b,a} = -D_{a,b}, \quad b * a = -a * b \qquad \text{for all} \quad a, b \text{ in } \mathfrak{A}_0,$$

the fact that $[R_x, R_y] = -[R_y, R_x]$ is in $\mathfrak{D}(\mathfrak{J})$, (3.75) and the obvious generalization of (4.50). We shall only begin the proof of

Theorem 4.13 (Tits). *The algebra* $\mathfrak{L}$ *in* (4.75) *is a Lie algebra. If K is the algebraic closure of F, then, corresponding to the four (alternative) composition algebras* $\mathfrak{A}$ *over K and the five indicated Jordan algebras* $\mathfrak{J}$ *over K (that is, Ke and the four central simple algebras* $\mathfrak{H}(K_3)$, $\mathfrak{H}(\mathfrak{J}_3)$, $\mathfrak{H}(\mathfrak{Q}_3)$, *and* $\mathfrak{H}(\mathfrak{C}_3)$), *the Lie algebras* $\mathfrak{L}_K$ *are given in the table:*

		Ke	$\mathfrak{H}(K_3)$	$\mathfrak{H}(\mathfrak{J}_3)$	$\mathfrak{H}(\mathfrak{Q}_3)$	$\mathfrak{H}(\mathfrak{C}_3)$
	Ku	0	$\mathfrak{sl}(2)$	$\mathfrak{sl}(3)$	$\mathfrak{sp}(6)$	F_4
(4.80)	$\mathfrak{J}$	0	$\mathfrak{sl}(3)$	$\mathfrak{sl}(3) \oplus \mathfrak{sl}(3)$	$\mathfrak{sl}(6)$	E_6
	$\mathfrak{Q}$	$\mathfrak{sl}(2)$	$\mathfrak{sp}(6)$	$\mathfrak{sl}(6)$	$\mathfrak{o}(12)$	E_7
	$\mathfrak{C}$	G_2	F_4	E_6	E_7	E_8

where $\mathfrak{sl}(n)$ *denotes the special linear Lie algebra* $\mathfrak{C}'(V)$, $\dim V = n$. *Hence* $\mathfrak{L}$ *is central simple, except for three cases:* $\dim \mathfrak{A} = \dim \mathfrak{J} = 1$; $\dim \mathfrak{A} = 2$, $\dim \mathfrak{J} = 1$; $\dim \mathfrak{A} = 2$, $\dim \mathfrak{J} = 9$.

Proof. We shall prove here only that $\mathfrak{L}$ is a Lie algebra, and that in each case the dimension of $\mathfrak{L}$ is that of the algebra indicated in the table (4.80).

Since $\mathfrak{D}(\mathfrak{A})$ and $\mathfrak{D}(\mathfrak{J})$ are Lie algebras, and since the Jacobi identity is multilinear, in order to prove that $\mathfrak{L}$ is a Lie algebra, it is sufficient to verify the Jacobi identity

$$[[A, B], C] + [[B, C], A] + [[C, A], B] = 0$$

in six cases:

(4.81) $A = D$ in $\mathfrak{D}(\mathfrak{A})$; $B = D'$ in $\mathfrak{D}(\mathfrak{A})$;

$$C = a \otimes x, \quad a \text{ in } \mathfrak{A}_0, \quad x \text{ in } \mathfrak{J}_0;$$

(4.82) $A = E$ in $\mathfrak{D}(\mathfrak{J})$; $B = E'$ in $\mathfrak{D}(\mathfrak{J})$;

$$C = a \otimes x, \quad a \text{ in } \mathfrak{A}_0, \quad x \text{ in } \mathfrak{J}_0;$$

(4.83) $A = D$ in $\mathfrak{D}(\mathfrak{A})$; $B = a \otimes x$,

$$C = b \otimes y; \quad a, b \text{ in } \mathfrak{A}_0; \quad x, y \text{ in } \mathfrak{J}_0;$$

(4.84) $A = E$ in $\mathfrak{D}(\mathfrak{J})$; $B = a \otimes x$,

$$C = b \otimes y; \quad a, b \text{ in } \mathfrak{A}_0; \quad x, y \text{ in } \mathfrak{J}_0;$$

(4.85) $A = D$ in $\mathfrak{D}(\mathfrak{A})$; $B = E$ in $\mathfrak{D}(\mathfrak{J})$;

$$C = a \otimes x, \quad a \text{ in } \mathfrak{A}_0, \quad x \text{ in } \mathfrak{J}_0;$$

(4.86) $A = a \otimes x, \quad B = b \otimes y$,

$$C = c \otimes z; \quad a, b, c \text{ in } \mathfrak{A}_0; \quad x, y, z \text{ in } \mathfrak{J}_0.$$

By (4.77) we have

$$[[D, D'], a \otimes x] + [[D', a \otimes x], D] + [[a \otimes x, D], D']$$

$$= -a[D, D'] \otimes x - aD'D \otimes x + aDD' \otimes x = 0,$$

verifying the case (4.81). The case (4.82) follows similarly from (4.78). The multiplication rules (4.76)–(4.79) next imply

$$[[D, a \otimes x], b \otimes y] + [[a \otimes x, b \otimes y], D] + [[b \otimes y, D], a \otimes x]$$

$$= -[aD \otimes x, b \otimes y] + \tfrac{1}{12} \langle x, y \rangle [D_{a,b}, D] + [(a * b) \otimes (x * y), D]$$

$$- (a, b)[[R_x, R_y], D] - [a \otimes x, bD \otimes y] = 0$$

by (3.72) and (4.62). For we have seen in Chapter III that D is skew with respect to (a, b) from which it follows also that $(a * b)D =$

$aD * b + a * bD$. We have verified case (4.83); similarly we have (4.84). Case (4.85) is given by

$$[[D, E], a \otimes x] + [[E, a \otimes x], D] + [[a \otimes x, D], E]$$

$$= -aD \otimes xE + aD \otimes xE = 0.$$

Now $[[a \otimes x, b \otimes y], c \otimes z] = \frac{1}{12}\langle xy, z\rangle D_{a * b,c} + (-\frac{1}{12}cD_{a,b} \otimes \langle x, y\rangle z + (a * b) * c \otimes (x * y) * z + (a, b)c \otimes (x, z, y)) - (ab, c)[R_{xy}, R_z]$ since $\langle x * y, z\rangle = \langle xy, z\rangle - \frac{1}{3}\langle x, y\rangle\langle e, z\rangle = \langle xy, z\rangle$ when $T(z) = 0$, and similarly $(a * b, c) = (ab, c)$ for c in $\mathfrak{A}_0$. Use $\sum$ to denote the sum of three terms obtained by cyclic permutation of a, b, c in $\mathfrak{A}_0$ and x, y, z in $\mathfrak{J}_0$. The case (4.86) will be verified when we show

$$\sum \langle xy, z\rangle D_{a * b,c} = 0,$$

$$\sum (ab, c)[R_{xy}, R_z] = 0,$$

and

$$(4.87) \qquad -\frac{1}{12}\sum cD_{a,b} \otimes \langle x, y\rangle z + \sum(a * b) * c \otimes (x * y) * z$$

$$+ \sum (a, b)c \otimes (x, z, y) = 0$$

for all a, b, c in $\mathfrak{A}_0$; x, y, z in $\mathfrak{J}_0$. Since $\langle x, y\rangle$ is a trace form on $\mathfrak{J}$, we have $\langle xy, z\rangle = \langle yz, x\rangle = \langle zx, y\rangle$, so that

$$\sum \langle xy, z\rangle D_{a * b,c} = \langle xy, z\rangle \sum D_{ab,c} = 0$$

by (3.73), since $D_{u,c} = 0$. Similarly $(ab, c) = (bc, a) = (ca, b)$ for all a, b, c in $\mathfrak{A}_0$, implying that

$$\sum (ab, c)[R_{xy}, R_z] = (ab, c) \sum [R_{xy}, R_z] = 0$$

by (4.4). Finally,

$$(x, z, y) = \frac{1}{3}(\langle xz, y\rangle - \langle x, zy\rangle)e + \frac{1}{3}\langle x, z\rangle y + (x * z) * y$$

$$- \frac{1}{3}\langle z, y\rangle x - x * (z * y),$$

so that

$$\sum (a, b)c \otimes (x, z, y) = \tfrac{1}{3} \sum (a, b)c \otimes \langle z, x \rangle y + \sum (a, b)c \otimes (z * x) * y$$
$$- \tfrac{1}{3} \sum (a, b)c \otimes \langle y, z \rangle x - \sum (a, b)c \otimes (y * z) * x$$
$$= \sum \{(b, c)a - (c, a)b\} \otimes \{\tfrac{1}{3}\langle x, y \rangle z + (x * y) * z\},$$

implying that (4.87) is equivalent to

$$(4.88) \quad \sum \{ -\tfrac{1}{4}cD_{a,b} + (b, c)a - (c, a)b \} \otimes \tfrac{1}{3}\langle x, y \rangle z$$
$$+ \sum \{ (b, c)a - (c, a)b + (a * b) * c \} \otimes (x * y) * z = 0$$

for all a, b, c in $\mathfrak{A}_0$; x, y, z in $\mathfrak{J}_0$. Equation (3.74) implies the generalization

$$[[a, b], c] + [[b, c], a] + [[c, a], b] = 6(a, b, c)$$

of the Jacobi identity to alternative algebras. We use this as follows for a, b, c in $\mathfrak{A}_0$: $b * a = -a * b$ implies $[a, b] = 2a * b$, $[[a, b], c] = 4(a * b) * c$, so that

$$\tfrac{3}{2}(a, b, c) = (a * b) * c + (b * c) * a - b * (a * c)$$
$$= (a * b) * c + (b, c, a) + (b, c)a$$
$$- (c, a)b + ((bc, a) - (b, ca))u,$$

implying

$$(4.89) \quad (b, c)a - (c, a)b + (a * b) * c = \tfrac{1}{2}(a, b\ c) \qquad \text{for all} \quad a, b\ c \text{ in } \mathfrak{A}_0.$$

Then

$$cD_{a,b} = [c, [a, b]] - 3(a, c, b) = -4(a * b) * c + 3(a, b, c),$$

implying that

$$-\tfrac{1}{4}cD_{a,b} + (b, c)a - (c, a)b = (a * b) * c - \tfrac{3}{4}(a, b, c) + (b, c)a - (c, a)b$$
$$= -\tfrac{1}{4}(a, b, c)$$

by (4.89). Hence (4.88) is equivalent to

$$-\tfrac{1}{4} \sum (a, b, c) \otimes \tfrac{1}{3}\langle x, y \rangle z + \tfrac{1}{2} \sum (a, b, c) \otimes (x * y) * z = 0,$$

or, since $(a, b, c) = (b, c, a) = (c, a, b)$, to

(4.90) $\qquad (a, b, c) \otimes \sum \{2(x * y) * z - \frac{1}{3}\langle x, y \rangle z\} = 0$

for all a, b, c in $\mathfrak{A}_0$; x, y, z in $\mathfrak{J}_0$. But (4.41) yields

$$x^3 - \tfrac{1}{2}T(x^2)x - \tfrac{1}{3}T(x^3)e = 0 \qquad \text{for all} \quad x \text{ in } \mathfrak{J}_0.$$

Linearization gives

$$2 \sum (xy)z - \sum \langle x, y \rangle z - \frac{2}{3} \sum \langle xy, z \rangle e = 0.$$

Now $(xy)z = \frac{1}{3}\langle xy, z \rangle e + \frac{1}{3}\langle x, y \rangle z + (x * y) * z$, so this becomes

$$\sum \{2(x * y) * z - \tfrac{1}{3}\langle x, y \rangle z\} = 0 \qquad (x, y, z \text{ in } \mathfrak{J}_0).$$

We have proved (4.90), verifying case (4.86), and completing the proof that $\mathfrak{L}$ is a Lie algebra.

We know that

$$\mathfrak{D}(Ku) = \mathfrak{D}(\mathfrak{J}) = 0, \qquad \mathfrak{D}(\mathfrak{Q}) \cong [\mathfrak{Q}, \mathfrak{Q}] \cong \mathfrak{sl}(2), \qquad \mathfrak{D}(\mathfrak{C}) = G_2.$$

Also $\mathfrak{D}(Ke) = 0$. Moreover,

$$\mathfrak{D}(\mathfrak{H}(K_3)) = \mathfrak{o}(3) \cong \mathfrak{sl}(2), \quad \mathfrak{D}(\mathfrak{H}(\mathfrak{J}_3)) = \mathfrak{sl}(3), \quad \mathfrak{D}(\mathfrak{H}(\mathfrak{Q}_3)) = \mathfrak{sp}(6)$$

(Jacobson and Jacobson [1]), and $\mathfrak{D}(\mathfrak{H}(\mathfrak{C}_3)) = F_4$ by Theorem 4.9. These known facts establish the validity of the first column and first row of table (4.80), and give us the numbers in the first column and first row of the corresponding table of dimensions for the algebras in (4.80):

		1	6	9	15	27
	1	0	3	8	21	52
(4.91)	2	0	8	16	35	78
	4	3	21	35	66	133
	8	14	52	78	133	248

But then the other dimensions in (4.91) follow directly from the formula

$$\dim \mathfrak{L} = \dim \mathfrak{D}(\mathfrak{A}) + (-1 + \dim \mathfrak{A})(-1 + \dim \mathfrak{J}) + \dim \mathfrak{D}(\mathfrak{J})$$

where in each row (resp. column) the number in the first column (resp. row) is $\dim \mathfrak{D}(\mathfrak{A})$ (resp. $\dim \mathfrak{D}(\mathfrak{J})$).

We note that the representation of E_6 over K which is given in

Theorem 4.12 corresponds to E_6 in the last column of (4.80). For $\mathfrak{Z} = Ku + Kv_1$ with $v_1{}^2 = u$ implies $\mathfrak{D}(\mathfrak{Z}) = 0$, $\mathfrak{Z}_0 = Kv_1$. Then $\mathfrak{L} = Kv_1 \otimes \mathfrak{J}_0 + \mathfrak{D}(\mathfrak{J})$ is isomorphic to $E_6 = R_0(\mathfrak{J}) + \mathfrak{D}(\mathfrak{J})$ under the linear transformation

$$v_1 \otimes x + E \leftrightarrow R_x + E \qquad (x \text{ in } \mathfrak{J}_0, \quad E \text{ in } \mathfrak{D}(\mathfrak{J})).$$

For

$$[v_1 \otimes x, E] = v_1 \otimes xE \leftrightarrow R_{xE} = [R_x, E],$$

while

$$[v_1 \otimes x, v_1 \otimes y] = -n(v_1)[R_x, R_y] = [R_x, R_y] \leftrightarrow [R_x, R_y]$$

by (4.79).

In announcing this theorem Tits states that every "real form" of an exceptional simple Lie algebra G_2, F_4, E_6, E_7, or E_8 may be obtained by this construction (Tits [4]). The geometric significance of the "magic square" consisting of the last four columns of the table is summarized in Freudenthal [4].

V

POWER-ASSOCIATIVE ALGEBRAS

1. THE PEIRCE DECOMPOSITION

We recall that an algebra $\mathfrak{A}$ over F is called *power-associative* in case the subalgebra $F[x]$ generated by any element x of $\mathfrak{A}$ is associative. We have seen that this is equivalent to defining, for any x in $\mathfrak{A}$,

$$x^1 = x, \qquad x^{i+1} = xx^i \qquad \text{for} \quad i = 1, 2, 3, \ldots,$$

and requiring

$$(5.1) \qquad x^i x^j = x^{i+j} \qquad \text{for} \quad i, j = 1, 2, 3, \ldots.$$

All algebras mentioned in the Introduction are power-associative (Lie algebras trivially, since $x^2 = 0$ implies $x^i = 0$ for $i = 2, 3, \ldots$). We shall encounter in this chapter new examples of power-associative algebras.

The most important tool in the study of noncommutative power-associative algebras $\mathfrak{A}$ is the passage to the commutative algebra $\mathfrak{A}^+$. *Let F have characteristic $\neq 2$ throughout Chapter V*; we shall also require that F contains at least four distinct elements. The algebra $\mathfrak{A}^+$ is the same vector space as $\mathfrak{A}$ over F, but multiplication in $\mathfrak{A}^+$ is defined by

$$x \cdot y = \tfrac{1}{2}(xy + yx) \qquad \text{for} \quad x, y \text{ in } \mathfrak{A},$$

where xy is the (nonassociative) product in $\mathfrak{A}$. If $\mathfrak{A}$ is power-associative, then (as in the Introduction) powers in $\mathfrak{A}$ and $\mathfrak{A}^+$ coincide, and it follows that $\mathfrak{A}^+$ is a commutative power-associative algebra.

Let $\mathfrak{A}$ be power-associative. Then (5.1) implies

(5.2) $$x^2x = xx^2 \qquad \text{for all} \quad x \text{ in } \mathfrak{A}$$

and

(5.3) $$x^2x^2 = x(xx^2) \qquad \text{for all} \quad x \text{ in } \mathfrak{A}.$$

In terms of associators, we have

(5.4) $$(x, x, x) = 0 \qquad \text{for all} \quad x \text{ in } \mathfrak{A}$$

and

(5.5) $$(x, x, x^2) = 0 \qquad \text{for all} \quad x \text{ in } \mathfrak{A}.$$

Also (5.2) may be written in terms of a commutator as

(5.6) $$[x^2, x] = 0 \qquad \text{for all} \quad x \text{ in } \mathfrak{A}.$$

Using the linearization process employed in Chapter IV, we obtain from (5.6), by way of the intermediate identity

(5.7) $$2[x \cdot y, x] + [x^2, y] = 0 \qquad \text{for all} \quad x, y \text{ in } \mathfrak{A},$$

the multilinear identity

(5.8) $$[x \cdot y, z] + [y \cdot z, x] + [z \cdot x, y] = 0 \qquad \text{for all} \quad x, y, z \text{ in } \mathfrak{A}.$$

Similarly, if we assume that there are four distinct elements in F, (5.5) is equivalent to

(5.9) $$2(x, x, x \cdot y) + (x, y, x^2) + (y, x, x^2) = 0 \qquad \text{for all} \quad x, y \text{ in } \mathfrak{A},$$

to

(5.10)
$$\begin{aligned}
&2(x, y, x \cdot z) + 2(x, z, x \cdot y) + 2(y, x, x \cdot z) + 2(x, x, y \cdot z) \\
&+ 2(z, x, x \cdot y) + (y, z, x^2) + (z, y, x^2) = 0
\end{aligned}$$

for all x, y, z in $\mathfrak{A}$, and finally to the multilinear identity

(5.11)
$$\begin{aligned}
&(x, y, z \cdot w) + (z, y, w \cdot x) + (w, y, x \cdot z) \\
&+ (y, x, z \cdot w) + (z, x, w \cdot y) + (w, x, y \cdot z) \\
&+ (z, w, x \cdot y) + (x, w, y \cdot z) + (y, w, z \cdot x) \\
&+ (w, z, x \cdot y) + (x, z, y \cdot w) + (y, z, w \cdot x) = 0
\end{aligned}$$

for all x, y, z, w in $\mathfrak{A}$, where in each row of the formula (5.11) we have left one of the four elements x, y, z, w fixed in the middle position of the associator and permuted the remaining three cyclically.

We omit the proof of the fact that, if F has characteristic 0, then identities (5.2) and (5.3) are sufficient to ensure that an algebra is power-associative; the proof involves inductions employing the multilinear identities (5.8) and (5.11). We omit similarly a proof of the fact that, if F has characteristic $\neq 2$, 3, 5, the single identity (5.3) in a commutative algebra implies power-associativity. One consequence of this latter fact is that in a number of proofs concerning power-associative algebras separate consideration has to be given to the characteristic 3 or 5 case by bringing in associativity of fifth or sixth powers and an assumption that F contains at least 6 distinct elements. We shall omit these details, simply by assuming characteristic $\neq 3$, 5 upon occasion.

An algebra $\mathfrak{A}$ over F is called *strictly power-associative* in case every scalar extension $\mathfrak{A}_K$ is power-associative. If $\mathfrak{A}$ is a commutative power-associative algebra over F of characteristic $\neq 2$, 3, 5, then $\mathfrak{A}$ is strictly power-associative. The assumption of strict power-associativity is employed in the noncommutative case, and in the commutative case of characteristic 3 or 5, when one wishes to use the method of extension of the base field.

Let $\mathfrak{A}$ be a finite-dimensional power-associative algebra over F. Just as in the proofs of Propositions 2.2 and 2.3, one may argue that $\mathfrak{A}$ has a unique maximal nilideal $\mathfrak{N}$, and that $\mathfrak{A}/\mathfrak{N}$ has maximal nilideal 0. For if $\mathfrak{A}$ is a power-associative algebra which contains a nilideal $\mathfrak{B}$ such that $\mathfrak{A}/\mathfrak{B}$ is a nilalgebra, then $\mathfrak{A}$ is a nilalgebra. [If x is in $\mathfrak{A}$, then $\overline{x^s} = \bar{x}^s = 0$ for some s, so that $x^s = y$ in $\mathfrak{B}$ and $x^{rs} = (x^s)^r = y^r = 0$ for some r.] Since any homomorphic image of a nilalgebra is a nilalgebra, it follows from the second isomorphism theorem that, if $\mathfrak{B}$ and $\mathfrak{C}$ are nilideals, then so is $\mathfrak{B} + \mathfrak{C}$. For $(\mathfrak{B} + \mathfrak{C})/\mathfrak{C} \cong \mathfrak{B}/(\mathfrak{B} \cap \mathfrak{C})$ is a nilalgebra, so $\mathfrak{B} + \mathfrak{C}$ is. This establishes the uniqueness of $\mathfrak{N}$. It follows as in the proof of Proposition 2.3 that 0 is the only nilideal of $\mathfrak{A}/\mathfrak{N}$. $\mathfrak{N}$ is called the *nilradical* of $\mathfrak{A}$, and $\mathfrak{A}$ is called *semisimple* in case $\mathfrak{N} = 0$. Of course any anticommutative algebra (for example, any Lie algebra) is a nilalgebra, and hence is its own nilradical. Hence this concept of semisimplicity is trivial for anticommutative algebras.

For the moment let $\mathfrak{A}$ be a commutative power-associative algebra, and let e be an idempotent in $\mathfrak{A}$. Putting $x = e$ in (5.9) and using commutativity, we have $y(2R_e^3 - 3R_e^2 + R_e) = 0$ for all y in $\mathfrak{A}$, or

(5.12) $$2R_e^3 - 3R_e^2 + R_e = 0$$

for any idempotent e in a commutative power-associative algebra $\mathfrak{A}$. As we have seen in the case of Jordan algebras in Chapter IV, this gives a Peirce decomposition

(5.13) $$\mathfrak{A} = \mathfrak{A}_1 + \mathfrak{A}_{1/2} + \mathfrak{A}_0$$

of $\mathfrak{A}$ as a vector space direct sum of subspaces $\mathfrak{A}_i$ defined by

$$\mathfrak{A}_i = \{x_i \mid x_i e = i x_i\}, \qquad i = 1, \tfrac{1}{2}, 0; \quad \mathfrak{A} \text{ commutative.}$$

Now if $\mathfrak{A}$ is any power-associative algebra, the algebra $\mathfrak{A}^+$ is a commutative power-associative algebra. Hence we have the Peirce decomposition (5.13) where

(5.14) $$\mathfrak{A}_i = \{x_i \mid e x_i + x_i e = 2 i x_i\}, \qquad i = 1, \tfrac{1}{2}, 0.$$

Put $x = e$ in (5.7) and let $y = x_i$ be in $\mathfrak{A}_i$ as in (5.14) to obtain $(2i - 1)[x_i, e] = 0$; that is, $x_i e = e x_i$ if $i \neq \tfrac{1}{2}$. Hence (5.14) becomes

(5.15)
$$\mathfrak{A}_i = \{x_i \mid e x_i = x_i e = i x_i\}, \qquad i = 1, 0;$$
$$\mathfrak{A}_{1/2} = \{x_{1/2} \mid e x_{1/2} + x_{1/2} e = x_{1/2}\}$$

in the Peirce decomposition (5.13) of any power-associative algebra $\mathfrak{A}$. As we have just seen, the properties of commutative power-associative algebras may be used (via $\mathfrak{A}^+$) to obtain properties of arbitrary power-associative algebras.

Let $\mathfrak{A}$ be a commutative power-associative algebra with Peirce decomposition (5.13) relative to an idempotent e. Then $\mathfrak{A}_1$ and $\mathfrak{A}_0$ are orthogonal subalgebras of $\mathfrak{A}$ which are related to $\mathfrak{A}_{1/2}$ as follows:

(5.16)
$$\mathfrak{A}_{1/2}\,\mathfrak{A}_{1/2} \subseteq \mathfrak{A}_1 + \mathfrak{A}_0,$$
$$\mathfrak{A}_1\,\mathfrak{A}_{1/2} \subseteq \mathfrak{A}_{1/2} + \mathfrak{A}_0,$$
$$\mathfrak{A}_0\,\mathfrak{A}_{1/2} \subseteq \mathfrak{A}_1 + \mathfrak{A}_{1/2}.$$

Note that the last two inclusion relations of (5.16) are weaker than for Jordan algebras in (4.16). The proofs are similar to those in Chapter IV, and are given by putting $x = e$, $y = y_j \in \mathfrak{A}_j$, $z = x_i \in \mathfrak{A}_i$ in (5.10). We omit the details except to note that the characteristic 3 case of orthogonality requires associativity of fifth powers.

For x in $\mathfrak{A}_i$, w in $\mathfrak{A}_{1/2}$, we have $wx = (wx)_{1/2} + (wx)_0$ in $\mathfrak{A}_{1/2} + \mathfrak{A}_0$ by (5.16). Then $w \rightarrow (wx)_{1/2}$ is a linear operator on $\mathfrak{A}_{1/2}$ which we denote by S_x:

$$wS_x = (wx)_{1/2} \quad \text{for} \quad x \text{ in } \mathfrak{A}_1, \quad w \text{ in } \mathfrak{A}_{1/2}.$$

If $\mathfrak{H}$ is the (associative) algebra of all linear operators on $\mathfrak{A}_{1/2}$, then $x \rightarrow 2S_x$ is a homomorphism of $\mathfrak{A}_1$ into the special Jordan algebra $\mathfrak{H}^+$. For $x \rightarrow S_x$ is clearly linear and we verify

(5.17) $$S_{xy} = S_x S_y + S_y S_x \quad \text{for all} \quad x, y \text{ in } \mathfrak{A}_1$$

as follows: put $x \in \mathfrak{A}_1$, $z = e$, $y = w \in \mathfrak{A}_{1/2}$ in (5.10) to obtain

(5.18) $$e[2(wx)x + wx^2] + wx^2 - 4(wx)x = 0,$$

since $e(xw) = \frac{1}{2}(xw)_{1/2}$ implies $x[e(xw)] = \frac{1}{2}x(xw)_{1/2} = \frac{1}{2}x(wx)$. By taking the $\mathfrak{A}_{1/2}$ component in (5.18), we have

$$S_{x^2} = 2S_x^{\,2}$$

after dividing by 3. Linearizing, we have (5.17). Similarly, defining the linear operator T_z on $\mathfrak{A}_{1/2}$ for any z in $\mathfrak{A}_0$ by

$$wT_z = (wz)_{1/2} \quad \text{for} \quad z \text{ in } \mathfrak{A}_0, \quad w \text{ in } \mathfrak{A}_{1/2},$$

we have

(5.19) $$T_{zy} = T_z T_y + T_y T_z \quad \text{for all} \quad z, y \text{ in } \mathfrak{A}_0,$$

and

(5.20) $$S_x T_z = T_z S_x \quad \text{for all} \quad x \text{ in } \mathfrak{A}_1, \quad z \text{ in } \mathfrak{A}_0.$$

This is part of the basic machinery used in developing the structure theory for commutative power-associative algebras. All simple commutative strictly power-associative algebras of characteristic $\neq 2$ which are not nilalgebras (that is, which are semisimple in the sense that the nilradical is 0) and which have degree >2 are Jordan algebras, whereas counterexamples of degree 2 and characteristic $p > 0$ are known; all semisimple commutative power-associative algebras of characteristic 0 are Jordan algebras (Albert [13, 17, 19]; Kokoris [1–3]; Oehmke [3]). These results have been extended by the same technique to flexible power-associative algebras (the conclusion being that $\mathfrak{A}^+$ is Jordan) (Oehmke [1]; Kleinfeld and Kokoris [1]).

2. FINITE POWER-ASSOCIATIVE DIVISION RINGS

We have already developed as much of the technique mentioned above as will be required in the proof of the following generalization of Wedderburn's theorem that every finite associative division ring is a field (Artin [2], p. 37; Albert [24], p. 62). In Chapter IV it was mentioned that there are no exceptional Jordan division algebras over finite fields, so that any exceptional central simple Jordan algebra over a finite field is actually reduced (Albert [20]). We assume this result (as well as Wedderburn's theorem) in the proof of:

Theorem 5.1 (Albert). *Let $\mathfrak{D}$ be a finite power-associative division ring of characteristic $\neq 2, 3, 5$. Then $\mathfrak{D}$ is a field.*

For the proof we require two lemmas. Characteristics 3 and 5 could be included if we assumed strict power-associativity.

Lemma 5.2. *If e and e' are orthogonal idempotents in a commutative power-associative algebra $\mathfrak{A}$, then*

$$(5.21) \qquad (e, a, e') = 0 \qquad \text{for all} \quad a \text{ in } \mathfrak{A}.$$

Proof. Let

$$\mathfrak{A} = \mathfrak{A}_1 + \mathfrak{A}_{1/2} + \mathfrak{A}_0$$

be the Peirce decomposition (5.13) of $\mathfrak{A}$ relative to the idempotent e. Relative to the idempotent e' we have also a Peirce decomposition, where we shall write $\mathfrak{A}_i'$ for $\mathfrak{A}_{i,e'}$ ($i = 1, \frac{1}{2}, 0$). Then

$$(5.22) \qquad \mathfrak{A} = \mathfrak{A}_1' + \mathfrak{A}_{1/2}' + \mathfrak{A}_0'.$$

Note that e is in $\mathfrak{A}_0'$, while e' is in $\mathfrak{A}_0$. We begin by showing that

$$(5.23) \qquad e'\mathfrak{A}_{1/2} \subseteq \mathfrak{A}_{1/2}.$$

Let w be in $\mathfrak{A}_{1/2}$. Since e' is in $\mathfrak{A}_0$, we have

$$e'w = (e'w)_1 + (e'w)_{1/2} = (e'w)_1 + wT_{e'},$$

where

$$(5.24) \qquad T_{e'} = 2T_{e'}^2$$

by (5.19). If $w = w_1' + w_{1/2}' + w_0'$ with w_i' in $\mathfrak{A}_i'$, then $w_1' = 2e'(e'w) - e'w$ is in $\mathfrak{A}_1$ by (5.24). Hence $w_1' = e'w_1' \in \mathfrak{A}_0 \mathfrak{A}_1 = 0$, and we have

(5.25) $$w = w_{1/2}' + w_0' \qquad (w_i' \text{ in } \mathfrak{A}_i').$$

Then $\tfrac{1}{2}w = ew = ew_{1/2}' + ew_0'$, where $ew_{1/2}' \in \mathfrak{A}_0' \mathfrak{A}_{1/2}' \subseteq \mathfrak{A}_1' + \mathfrak{A}_{1/2}'$ and $ew_0' \in \mathfrak{A}_0' \mathfrak{A}_0' \subseteq \mathfrak{A}_0'$. It follows that

(5.26) $$\tfrac{1}{2}w_{1/2}' = ew_{1/2}',$$

so that $w_{1/2}'$ is in $\mathfrak{A}_{1/2}$. Thus $e'w = \tfrac{1}{2}w_{1/2}'$ is in $\mathfrak{A}_{1/2}$, and we have (5.23). If $a = a_1 + a_{1/2} + a_0$ $(a_i \in \mathfrak{A}_i)$ is any element of $\mathfrak{A}$, then

$$(e, a, e') = (e, a_1, e') + (e, a_{1/2}, e') + (e, a_0, e')$$
$$= (ea_{1/2})e' - e(a_{1/2}e') = \tfrac{1}{2}a_{1/2}e' - \tfrac{1}{2}a_{1/2}e' = 0$$

by (5.23).

Lemma 5.3. *Any finite-dimensional power-associative division algebra $\mathfrak{D}$ has an identity element* 1.

Proof. The division algebra $\mathfrak{D}$ is without nilpotent elements $\neq 0$. Hence $\mathfrak{D}$ contains an idempotent e by Proposition 3.3. Let

(5.27) $$\mathfrak{D} = \mathfrak{D}_1 + \mathfrak{D}_{1/2} + \mathfrak{D}_0$$

be the Peirce decomposition of $\mathfrak{D}$ relative to e which is given by (5.13) and (5.15). Since $e\mathfrak{D}_0 = 0$ in the division algebra $\mathfrak{D}$, we have $\mathfrak{D}_0 = 0$. For any y in $\mathfrak{D}_{1/2}$, we have $y^2 = y \cdot y$ in $\mathfrak{D}_1 + \mathfrak{D}_0 = \mathfrak{D}_1$ by (5.16). Also $2e \cdot y = ey + ye = y$ implies

$$0 = 2(y, y, e \cdot y) + (y, e, y^2) + (e, y, y^2)$$
$$= (ye)y^2 - y(ey^2) + (ey)y^2 - ey^3$$
$$= y^3 - y^3 - ey^3 = -ey^3$$

by (5.9). Then $y = 0$, since $\mathfrak{D}$ is a division algebra. Thus $\mathfrak{D}_{1/2} = 0$, $\mathfrak{D} = \mathfrak{D}_1$, e is the identity element in $\mathfrak{D}$.

Proof of Theorem 5.1. We are assuming that $\mathfrak{D}$ is a finite power-associative division ring. We have seen in Chapter II that this means that $\mathfrak{D}$ is a (finite-dimensional) division algebra over a (finite) field. Hence Lemma 5.3 implies that $\mathfrak{D}$ has a unity element 1, so that $\mathfrak{D}$ is an algebra over its center. Thus we may as well take $\mathfrak{D}$ to be an algebra over its center F, a finite field. Hence F is perfect (Zariski and Samuel [1], p. 65).

Now $\mathfrak{D}^+$ is a Jordan algebra over F. For let x, y be any elements of $\mathfrak{D}^+$. If x is in $F1$, the Jordan identity

$$(x \cdot y) \cdot x^2 = x \cdot (y \cdot x^2) \qquad \text{for all} \quad x, y \text{ in } \mathfrak{D}^+$$

holds trivially. Otherwise the (commutative associative) subalgebra $F[x]$ of $\mathfrak{D}^+$ is a finite (necessarily separable) extension of F, so there is an extension K of F such that $F[x]_K = K \oplus K \oplus \cdots \oplus K$, x is a linear combination $x = \xi_1 e_1 + \xi_2 e_2 + \cdots + \xi_n e_n$ of pairwise orthogonal idempotents e_i in $F[x]_K \subseteq (\mathfrak{D}^+)_K$ with coefficients in K. In order to establish the Jordan identity in $\mathfrak{D}^+$, it is sufficient to establish

$$(5.28) \quad (e_i \cdot y) \cdot (e_j \cdot e_k) = e_i \cdot [y \cdot (e_j \cdot e_k)], \qquad i, j, k = 1, \dots, n.$$

For $j \neq k$, this is obvious; for $j = k$, (5.28) reduces to Lemma 5.2.

Now the radical of $\mathfrak{D}^+$ (consisting of nilpotent elements) is 0. Although our proof of Corollary 4.6 is valid only for characteristic 0, we remarked in Chapter IV that the conclusion is valid for characteristic $\neq 2$. Hence $\mathfrak{D}^+$ is a direct sum $\mathfrak{S}_1 \oplus \cdots \oplus \mathfrak{S}_r$ of r simple ideals $\mathfrak{S}_i$, each with unity element e_i. The existence of an idempotent $e \neq 1$ in $\mathfrak{D}^+$ is sufficient to give zero divisors in $\mathfrak{D}$, a contradiction, since the product $e(1 - e) = 0$ in $\mathfrak{D}$. Hence $r = 1$ and $\mathfrak{D}^+$ is a simple Jordan algebra over F. Let C be the center of $\mathfrak{D}^+$. Then C is a finite separable extension of F, $C = F[z]$, z in C (Zariski and Samuel [1], p. 84; Artin [1], p. 66). If $\mathfrak{D}^+ = C = F[z]$, then $\mathfrak{D} = F[z]$ is a field, and the theorem is established. Hence we may assume that $\mathfrak{D}^+ \neq C$, so $\mathfrak{D}^+$ is a central simple Jordan algebra of degree $t \geq 2$ over the finite field C and is of one of the types A–E listed in Chapter IV. We are assuming that any algebra of type E is known to be a reduced algebra. Then $\mathfrak{D}$ contains an idempotent $e_1 \neq 1$, a contradiction. The other types are eliminated as follows.

Wedderburn's theorem implies that, over any finite field, there are no associative central division algebras of dimension > 1. Hence, by Wedderburn's theorem on simple associative algebras, every associative central simple algebra over a finite field is a total matrix algebra. Thus we have the following possibilities:

A_I. $\mathfrak{D}^+ \cong C_t^{\,+}$, $t \geq 2$. Then $C_t^{\,+}$ contains an idempotent $e_{11} \neq 1$, a contradiction.

A_{II}. $\mathfrak{D}^+$ is the set $\mathfrak{H}(\mathfrak{Z}_t)$ of self-adjoint elements in $\mathfrak{Z}_t$, $\mathfrak{Z}$ a quadratic extension of C, where the involution may be taken to be $a \to g^{-1}\bar{a}'g$ with g a diagonal matrix. Hence $\mathfrak{H}(\mathfrak{Z}_t)$ contains $e_{11} \neq 1$, a contradiction.

B. $\mathfrak{D}^+ \cong \mathfrak{H}(C_t)$, the involution being $a \to g^{-1}a'g$ with g diagonal; hence $\mathfrak{H}(C_t)$ contains $e_{11} \neq 1$, a contradiction.

C. $\mathfrak{D}^+ \cong \mathfrak{H}(C_{2t})$, the involution being

$$a \to g^{-1}a'g, \qquad g = \begin{pmatrix} 0 & 1_t \\ -1_t & 0 \end{pmatrix};$$

$\mathfrak{H}(C_{2t})$ contains the idempotent $e_{11} + e_{t+1,t+1} \neq 1$, a contradiction.

D. $\mathfrak{D}^+ = F1 + \mathfrak{M}$, dim $\mathfrak{M} \geq 2$. If dim $\mathfrak{M} = 2$, we have $\mathfrak{D}^+ \cong \mathfrak{H}(C_2)$ of type B above. Otherwise dim $\mathfrak{M} \geq 3$, and there are isotropic vectors $x \neq 0$ in $\mathfrak{M}$ satisfying $x^2 = (x, x)1 = 0$, a contradiction (Artin [2], p. 144).

See McCrimmon [5] for a proof of Theorem 5.1 which depends on the methods of Shirshov and Cohn mentioned on page 93, rather than on the classification of finite-dimensional central simple Jordan algebras.

3. NONCOMMUTATIVE JORDAN ALGEBRAS

When trace forms exist on finite-dimensional power-associative algebras, the results of imposing natural conditions on idempotent and nilpotent elements are indeed striking.

Theorem 5.4 (Albert).   *Let $\mathfrak{A}$ be a finite-dimensional power-associative algebra over F satisfying the following conditions:*

(i)   *there is an (associative) trace form (x, y) defined on $\mathfrak{A}$;*

(ii)   *$(e, e) \neq 0$ for any idempotent e in $\mathfrak{A}$;*

(iii)   *$(x, y) = 0$ if $x \cdot y$ is nilpotent, x, y in $\mathfrak{A}$.*

Then the nilradical $\mathfrak{N}$ of $\mathfrak{A}$ coincides with the nilradical of $\mathfrak{A}^+$, and is the radical $\mathfrak{A}^\perp$ of the trace form (x, y). The semisimple power-associative *algebra* $\mathfrak{S} = \mathfrak{A}/\mathfrak{N}$ *satisfies* (i)–(iii) *with* (x, y) *nondegenerate. For any such* $\mathfrak{S}$ *we have*

(a) $\mathfrak{S} = \mathfrak{S}_1 \oplus \cdots \oplus \mathfrak{S}_t$ *for simple* $\mathfrak{S}_i$;

(b) $\mathfrak{S}$ *is flexible.*

If F has characteristic $\neq 5$, *then*

(c) $\mathfrak{S}^+$ *is a semisimple Jordan algebra*;

(d) $\mathfrak{S}_i{}^+$ *is a simple (Jordan) algebra,* $i = 1, ..., t$.

Proof. By (i) we know from Chapter II that $\mathfrak{A}^\perp$ is an ideal of $\mathfrak{A}$. If there were an idempotent e in $\mathfrak{A}^\perp$, then (ii) would imply $(e, e) \neq 0$, a contradiction. Hence $\mathfrak{A}^\perp$ is a nilideal: $\mathfrak{A}^\perp \subseteq \mathfrak{N}$. Conversely, x in $\mathfrak{N}$ implies $x \cdot y$ is in $\mathfrak{N}$ for all y in $\mathfrak{A}$, so that $(x, y) = 0$ for all y in $\mathfrak{N}$ by (iii), or x is in $\mathfrak{A}^\perp$. Hence $\mathfrak{N} \subseteq \mathfrak{A}^\perp$, $\mathfrak{N} = \mathfrak{A}^\perp$. Any ideal of $\mathfrak{A}$ is clearly an ideal of $\mathfrak{A}^+$; hence any nilideal of $\mathfrak{A}$ is a nilideal of $\mathfrak{A}^+$, and $\mathfrak{N}$ is contained in the nilradical $\mathfrak{N}_1$ of $\mathfrak{A}^+$. But x in $\mathfrak{N}_1$ implies $x \cdot y$ is in $\mathfrak{N}_1$ for all y in $\mathfrak{A}^+$, or $(x, y) = 0$ by (iii) and we have $\mathfrak{N}_1 \subseteq \mathfrak{A}^\perp = \mathfrak{N}$.

Now (x, y) induces a nondegenerate symmetric bilinear form $(\bar{x}, \bar{y})$ on $\mathfrak{A}/\mathfrak{A}^\perp = \mathfrak{A}/\mathfrak{N}$ where $\bar{x} = x + \mathfrak{N}$, etc.; that is, $(\bar{x}, \bar{y}) = (x, y)$. Then

$$(\bar{x}\bar{y}, \bar{z}) = (\overline{xy}, \bar{z}) = (xy, z) = (x, yz) = (\bar{x}, \bar{y}\bar{z}),$$

so $(\bar{x}, \bar{y})$ is a trace form. To show (ii) we take any idempotent $\bar{e}$ in $\mathfrak{A}/\mathfrak{N}$ and use the power-associativity of $\mathfrak{A}$ to "lift" the idempotent to $\mathfrak{A}$: $F[e]$ is a subalgebra of $\mathfrak{A}$ which is not a nilalgebra, so there is an idempotent u in $F[e] \subseteq Fe + \mathfrak{N}$, and $\bar{u} = \bar{e}$. Then $(\bar{e}, \bar{e}) = (\bar{u}, \bar{u}) = (u, u) \neq 0$. Suppose $\bar{x} \cdot \bar{y} = \overline{x \cdot y}$ is nilpotent. Then some power of $x \cdot y$ is in $\mathfrak{N}$, $x \cdot y$ is nilpotent, and $(\bar{x}, \bar{y}) = (x, y) = 0$, establishing (iii).

Now let $\mathfrak{S}$ satisfy (i)–(iii) with (x, y) nondegenerate. Then the nilradical of $\mathfrak{S}$ is 0, and the hypotheses of Theorem 2.6 apply. For if $\mathfrak{B}^2 = 0$ for an ideal $\mathfrak{B}$ of $\mathfrak{S}$, then $\mathfrak{B}$ is a nilideal, $\mathfrak{B} = 0$. We have $\mathfrak{S} = \mathfrak{S}_1 \oplus \cdots \oplus \mathfrak{S}_t$ for simple $\mathfrak{S}_i$; also we know that the $\mathfrak{S}_i$ are not nilalgebras (for then they would be nilideals of $\mathfrak{S}$), but this will follow from (d).

Now (5.7) implies that

$$0 = (2[x \cdot y, x] + [x^2, y], z)$$

$$= ((xy)x, z) + ((yx)x, z) - (x(xy), z)$$

$$- (x(yx), z) + (x^2y, z) - (yx^2, z)$$

for all x, y, z in $\mathfrak{S}$. The properties of a trace form imply that

(5.29) $(xy + yx, xz - zx) = (x^2, zy - yz)$.

Interchange z and y in (5.29) to obtain

$$(xz + zx, xy - yx) = (x^2, yz - zy) = (xy + yx, zx - xz).$$

Add $(xy + yx, xz + zx)$ to both sides of this to obtain $(xy, xz + zx) = (xy + yx, zx)$. Then $(xy, xz) = (yx, zx)$ so that

(5.30) $((xy)x, z) = (x(yx), z)$ for all x, y, z in $\mathfrak{S}$.

Since (x, y) is nondegenerate on $\mathfrak{S}$, $(xy)x = x(yx)$; that is, $\mathfrak{S}$ is flexible.
   To prove (c) we note first that (x, y) is a trace form on $\mathfrak{S}^+$:

(5.31) $(x \cdot y, z) = (x, y \cdot z)$ for all x, y, z in $\mathfrak{S}$.

Also it follows from (5.31), just as in formula (2.12), that

(5.32) $(yS_1 S_2 \cdots S_h, z) = (y, zS_h \cdots S_2 S_1)$

where S_i are right multiplications of the commutative algebra $\mathfrak{S}^+$. In the commutative power-associative algebra $\mathfrak{S}^+$ formula (5.9) becomes

(5.33) $4x^2 \cdot (x \cdot y) - 2x \cdot [x \cdot (x \cdot y)] - x \cdot (y \cdot x^2) - y \cdot x^3 = 0$.

Applying the same procedure as above, for all z in $\mathfrak{S}^+$ we have $4(x^2 \cdot (x \cdot y), z) - 2(x \cdot [x \cdot (x \cdot y)], z) - (x \cdot (y \cdot x^2), z) - (y \cdot x^3, z) = 0$ by (5.33), or

(5.34) $(y \cdot z, x^3) + 2(x \cdot [x \cdot (x \cdot y)], z) = 4(x \cdot y, x^2 \cdot z) - (y \cdot x^2, x \cdot z)$.

By (5.32) the left-hand side of (5.34) is unaltered by interchange of y and z. Hence

$$4(x \cdot y, x^2 \cdot z) - (y \cdot x^2, x \cdot z) = 4(x \cdot z, x^2 \cdot y) - (z \cdot x^2, x \cdot y)$$

so that, after dividing by 5, we have $(x \cdot y, x^2 \cdot z) = (y \cdot x^2, x \cdot z)$. Hence $((x \cdot y) \cdot x^2, z) = (x \cdot (y \cdot x^2), z)$ for all z in $\mathfrak{S}$, or $(x \cdot y) \cdot x^2 = x \cdot (y \cdot x^2)$, $\mathfrak{S}^+$ is a Jordan algebra. We know from Chapter IV that, since the nilradical of $\mathfrak{S}^+$ is 0, $\mathfrak{S}^+$ is a direct sum of simple ideals, but it is conceivable that these are not the $\mathfrak{S}_i^+$ given by (a). To see that the simple components of $\mathfrak{S}^+$ are the $\mathfrak{S}_i{}^+$ given by (a), we need to establish (d).

Let $\mathfrak{X}$ be an ideal of $\mathfrak{S}_i{}^+$; we need to show that $\mathfrak{X}$ is an ideal of $\mathfrak{S}_i$. It follows from (a) that $\mathfrak{X}$ is an ideal of $\mathfrak{S}^+$, and is therefore by (c) a direct sum of simple ideals of $\mathfrak{S}^+$ each of which has a unity element. The sum of these pairwise orthogonal idempotents in $\mathfrak{S}^+$ is the unity element e of $\mathfrak{X}$. Now e is an idempotent in $\mathfrak{S}^+$ (and $\mathfrak{S}$), and the Peirce decomposition (5.15) characterizes $\mathfrak{X}$ as

$$\mathfrak{X} = \mathfrak{S}_{1,e} = \{t \text{ in } \mathfrak{S} \mid et = te = t\}.$$

Now $\mathfrak{S}_{1/2,e} = \{s \text{ in } \mathfrak{S} \mid es + se = s\} = e \cdot \mathfrak{S}_{1/2,e} \subseteq \mathfrak{X} = \mathfrak{S}_{1,e}$ since $\mathfrak{X}$ is an ideal of $\mathfrak{S}^+$. Hence $\mathfrak{S}_{1/2,e} = 0$, $\mathfrak{S} = \mathfrak{S}_{1,e} \oplus \mathfrak{S}_{0,e}$ since $\mathfrak{S}_{1,e}$ and $\mathfrak{S}_{0,e}$ are orthogonal subalgebras. Hence $\mathfrak{X} = \mathfrak{S}_{1,e}$ is an ideal of $\mathfrak{S}$. Then $\mathfrak{X} \subseteq \mathfrak{S}_i$ is an ideal of $\mathfrak{S}_i$. Hence the only ideals of $\mathfrak{S}_i{}^+$ are 0 and $\mathfrak{S}_i{}^+$. Since $\mathfrak{S}_i{}^+$ cannot be a zero algebra, $\mathfrak{S}_i{}^+$ is simple.

We list without proof the central simple flexible algebras $\mathfrak{A}$ over F which are such that $\mathfrak{A}^+$ is a (central) simple Jordan algebra. These are the algebras which (over their centers) can appear as the simple components $\mathfrak{S}_i$ in (a) above:

1. $\mathfrak{A}$ is a central simple (commutative) Jordan algebra.

2. $\mathfrak{A}$ is a *quasiassociative* central simple algebra. That is, there is a scalar extension $\mathfrak{A}_K$, K a quadratic extension of F, such that $\mathfrak{A}_K$ is isomorphic to an algebra $\mathfrak{B}(\lambda)$ defined as follows: $\mathfrak{B}$ is a central simple associative algebra over K, $\lambda \neq \frac{1}{2}$ is a fixed element of K, and $\mathfrak{B}(\lambda)$ is the same vector space over K as $\mathfrak{B}$ but multiplication in $\mathfrak{B}(\lambda)$ is defined by

$$x * y = \lambda xy + (1 - \lambda)yx \qquad \text{for all} \quad x, y \text{ in } \mathfrak{B}$$

where xy is the (associative) product in $\mathfrak{B}$.

3. $\mathfrak{A}$ is a flexible quadratic algebra over F with nondegenerate norm form.

Note that, except for Lie algebras, all of the central simple algebras which we have mentioned in this book are listed here. Associative algebras are the case $\lambda = 1$ (also $\lambda = 0$) in 2; Cayley algebras are included in 3.

We should remark that, if an algebra $\mathfrak{A}$ contains 1, any trace form (x, y) on $\mathfrak{A}$ may be expressed in terms of a linear form $T(x)$. That is, we write

(5.35) $\qquad\qquad T(x) = (1, x) \qquad \text{for all} \quad x \text{ in } \mathfrak{A},$

and have

$$(x, y) = T(xy) \qquad \text{for all} \quad x, y \text{ in } \mathfrak{A}$$

since $(x1, y) = (1, xy)$. The symmetry and the associativity of the trace form (x, y) are equivalent to the vanishing of $T(x)$ on commutators and associators:

$$T(xy) = T(yx), \qquad T((xy)z) = T(x(yz)) \qquad \text{for all} \quad x, y, z \text{ in } \mathfrak{A}.$$

If $\mathfrak{A}$ is power-associative, hypotheses (ii) and (iii) of Theorem 5.4 become

$$T(e) \neq 0 \qquad \text{for any idempotent} \quad e \text{ in } \mathfrak{A},$$

and

(5.36) $$T(z) = 0 \qquad \text{for any nilpotent } z \text{ in } \mathfrak{A},$$

the latter being evident as follows: (5.36) implies that, if $x \cdot y$ is nilpotent, then

$$0 = T(x \cdot y) = (1, x \cdot y) = \tfrac{1}{2}(1, xy) + \tfrac{1}{2}(1, yx)$$
$$= \tfrac{1}{2}(x, y) + \tfrac{1}{2}(y, x) = (x, y)$$

and, conversely, if $z = 1 \cdot z$ is nilpotent, then (iii) implies $T(z) = (1, z) = 0$.

A natural generalization to noncommutative algebras of the class of (commutative) Jordan algebras is the class of algebras $\mathfrak{J}$ satisfying the Jordan identity

(5.37) $$(xy)x^2 = x(yx^2) \qquad \text{for all} \quad x, y \text{ in } \mathfrak{J}.$$

As in Chapter IV, we can linearize (5.37) to obtain

(5.38) $$(x, y, w \cdot z) + (w, y, z \cdot x) + (z, y, x \cdot w) = 0$$

for all w, x, y, z in $\mathfrak{J}$. If $\mathfrak{J}$ contains 1, then $w = 1$ in (5.38) implies

(5.39) $$(x, y, z) + (z, y, x) = 0 \qquad \text{for all} \quad x, y, z \text{ in } \mathfrak{J};$$

that is, $\mathfrak{J}$ is *flexible*:

(5.40) $$(xy)x = x(yx) \qquad \text{for all} \quad x, y \text{ in } \mathfrak{J}.$$

If a unity element 1 is adjoined to $\mathfrak{J}$ as in Chapter II, then a necessary and sufficient condition that the Jordan identity be satisfied in the algebra with 1 adjoined is that both the Jordan identity and the flexible law be satisfied in $\mathfrak{J}$. Hence we define a *noncommutative Jordan algebra* to be an algebra satisfying both (5.37) and (5.40).

It is easy to see that a flexible algebra $\mathfrak{J}$ is a noncommutative Jordan algebra if and only if any one of the following four statements is satisfied:

$$(x^2y)x = x^2(yx) \qquad \text{for all} \quad x, y \text{ in } \mathfrak{J};$$

$$x^2(xy) = x(x^2y) \qquad \text{for all} \quad x, y \text{ in } \mathfrak{J};$$

$$(yx)x^2 = (yx^2)x \qquad \text{for all} \quad x, y \text{ in } \mathfrak{J};$$

$\mathfrak{J}^+$ is a (commutative) Jordan algebra.

We leave these verifications as an exercise for the reader. We see from the last of these statements that any semisimple algebra (of characteristic $\neq 5$) satisfying the hypotheses of Theorem 5.4 is a noncommutative Jordan algebra.

Since (5.38) and (5.39) are multilinear, any scalar extension $\mathfrak{J}_K$ of a noncommutative Jordan algebra is a noncommutative Jordan algebra. It may be verified directly that any noncommutative Jordan algebra is power-associative (hence strictly power-associative).

Let $\mathfrak{J}$ be any noncommutative Jordan algebra. Then $\mathfrak{J}^+$ is a (commutative) Jordan algebra, and we have seen in Chapter IV that a trace form on $\mathfrak{J}^+$ may be given in terms of right multiplications of $\mathfrak{J}^+$. Our application of this to the situation here works more smoothly if there is a unity element 1 in $\mathfrak{J}$, so (if necessary) we adjoin one to $\mathfrak{J}$ to obtain a noncommutative Jordan algebra $\mathfrak{J}_1$ with 1 and having $\mathfrak{J}$ as a subalgebra (actually an ideal). Then by the proof of Theorem 4.5 we know that

$$(5.41) \quad (x, y) = \text{trace } R^+_{x \cdot y} = \tfrac{1}{2} \text{trace } (R_{x \cdot y} + L_{x \cdot y}) \qquad \text{for all} \quad x, y \text{ in } \mathfrak{J}_1$$

is a trace form on $\mathfrak{J}_1{}^+$ where R^+ indicates the right multiplication in $\mathfrak{J}_1{}^+$; hence (5.31) holds for all x, y, z in $\mathfrak{J}_1$ where (x, y) is the symmetric bilinear form (5.41). In terms of $T(x)$ defined by (5.35), we see that (5.31) is equivalent to

$$T((x \cdot y) \cdot z) = T(x \cdot (y \cdot z)) \qquad \text{for all} \quad x, y, z \text{ in } \mathfrak{J}_1.$$

Now (5.39) implies

$$L_{xy} - L_y L_x + R_y R_x - R_{yx} = 0 \qquad \text{for all} \quad x, y \text{ in } \mathfrak{J}_1.$$

Interchanging x and y, and subtracting, we have

$$R_{[x,y]} + L_{[x,y]} = [R_x, R_y] + [L_y, L_x] \qquad \text{for all} \quad x, y \text{ in } \mathfrak{J}_1.$$

Hence

$$T([x, y]) = (1, [x, y]) = \tfrac{1}{2}\, \text{trace}(R_{[x,y]} + L_{[x,y]}) = 0.$$

Then $xy = x \cdot y + \tfrac{1}{2}[x, y]$ implies

$$T(xy) = T(x \cdot y) = \tfrac{1}{2}T(xy) + \tfrac{1}{2}T(yx),$$

or

$$T(xy) = T(yx) = (x, y) \qquad \text{for all} \quad x, y \text{ in } \mathfrak{J}_1$$

since $T(x \cdot y) = (1, x \cdot y) = (x, y)$ by (5.31). Now (x, y) is a trace form on $\mathfrak{J}_1$. For

$$0 = 4T[(x \cdot y) \cdot z - x \cdot (y \cdot z)]$$

$$= T[(xy)z + (yx)z + z(xy) + z(yx) - x(yz) - x(zy) - (yz)x - (zy)x]$$

$$= 2T[(xy)z - x(yz) - (zy)x + z(yx)]$$

$$= 4T[(xy)z - x(yz)]$$

by (5.39), so $T((xy)z) = T(x(yz))$, or $(xy, z) = (x, yz)$, as desired. Then (5.41) induces a trace form on the subalgebra $\mathfrak{J}$ of $\mathfrak{J}_1$.

Corollary 5.5. *Modulo its nilradical, any finite-dimensional non-commutative Jordan algebra of characteristic 0 is (uniquely) expressible as a direct sum $\mathfrak{S}_1 \oplus \cdots \oplus \mathfrak{S}_t$ of simple ideals $\mathfrak{S}_i$. Over their centers these $\mathfrak{S}_i$ are central simple algebras of the following types: (commutative) Jordan, quasiassociative, or flexible quadratic.*

Proof. In order to apply Theorem 5.4 there remains only the verification of hypotheses (ii) and (iii) for (x, y) in (5.41). But these are immediate consequences of (4.14) and Proposition 4.4.

It was remarked in Chapter IV that, although proof was given only for commutative Jordan algebras of characteristic 0, the results were valid for arbitrary characteristic ($\neq 2$). The same statement cannot be made here. The trace argument in Theorem 5.4 can be modified to give the direct sum decomposition for semisimple algebras. But new central algebras occur for characteristic p; central simple algebras which are not listed in Corollary 5.5 are necessarily of degree one (Oehmke [1]; Kokoris [4, 5]).

A finite-dimensional power-associative algebra $\mathfrak{A}$ with 1 over F is called a *nodal algebra* in case every element of $\mathfrak{A}$ is of the form $\alpha 1 + z$ where α is in F and z is nilpotent, and $\mathfrak{A}$ is not of the form $\mathfrak{A} = F1 + \mathfrak{N}$ for $\mathfrak{N}$ a nil subalgebra of $\mathfrak{A}$. There are no such algebras which are alternative (of arbitrary characteristic), commutative Jordan (of characteristic $\neq 2$) (Jacobson [13]; McCrimmon [1]) or noncommutative Jordan of characteristic 0. But nodal noncommutative Jordan algebras of characteristic $p > 0$ do exist. Any nodal algebra has a homomorphic image which is a simple nodal algebra.

Let $\mathfrak{J}$ be a nodal noncommutative Jordan algebra over F. Since the commutative Jordan algebra $\mathfrak{J}^+$ is not a nodal algebra, $\mathfrak{J}^+ = F1 + \mathfrak{N}^+$ where $\mathfrak{N}^+$ is a nil subalgebra of $\mathfrak{J}^+$; that is, $\mathfrak{J} = F1 + \mathfrak{N}$, where $\mathfrak{N}$ is a subspace of $\mathfrak{J}$ consisting of all nilpotent elements of $\mathfrak{J}$, and $x \cdot y$ is in $\mathfrak{N}$ for all x, y in $\mathfrak{N}$. For any elements x, y in $\mathfrak{N}$ we have

$$(5.42) \qquad xy = \lambda 1 + z, \qquad \lambda \text{ in } F, \quad z \text{ in } \mathfrak{N}.$$

There must exist x, y in $\mathfrak{N}$ with $\lambda \neq 0$ in (5.42). Now (5.42) implies

$$yx = -\lambda 1 + (2x \cdot y - z)$$

and

$$(xy)x = \lambda x + zx = x(yx) = -\lambda x + 2x(x \cdot y) - xz,$$

or

$$x(x \cdot y) = \lambda x + x \cdot z.$$

Now

$$0 = (x, x, y) + (y, x, x)$$
$$= x^2 y - x(\lambda 1 + z) + (-\lambda 1 + 2x \cdot y - z)x - yx^2$$
$$= 2x^2 y - 2\lambda x - 2x \cdot z + 4(x \cdot y) \cdot x - 2x(x \cdot y) - 2x^2 \cdot y$$

implies

$$(5.43) \qquad x^2 y = 2\lambda x + 2x \cdot z - 2(x \cdot y) \cdot x + x^2 \cdot y.$$

Defining

$$x_i y = \lambda_i 1 + z_i, \qquad \lambda_i \text{ in } F, \quad z_i \text{ in } \mathfrak{N},$$

linearization of (5.43) gives

$$(x_1 \cdot x_2)y = \lambda_1 x_2 + \lambda_2 x_1 + x_1 \cdot z_2 + x_2 \cdot z_1 - (x_1 \cdot y) \cdot x_2$$

(5.44) $$\qquad\qquad - (x_2 \cdot y) \cdot x_1 + (x_1 \cdot x_2) \cdot y.$$

Theorem 5.6 (Kokoris). *Let $\mathfrak{J}$ be a simple nodal noncommutative Jordan algebra over F. Then F has characteristic p, $\mathfrak{J}^+$ is the p^n-dimensional (commutative) associative algebra $\mathfrak{J}^+ = F[1, x_1, \ldots, x_n], x_i{}^p = 0$, $n \geq 2$, and multiplication in $\mathfrak{J}$ is given by*

(5.45) $$fg = f \cdot g + \sum_{i,j=1}^{n} \frac{\partial f}{\partial x_i} \cdot \frac{\partial g}{\partial x_j} \cdot c_{ij}, \qquad c_{ij} = -c_{ji},$$

where at least one of the c_{ij} $(= -c_{ji})$ has an inverse.

Proof. Since $\mathfrak{J} = F1 + \mathfrak{N}$, every element a in $\mathfrak{J}$ is of the form

(5.46) $$a = \alpha 1 + x, \qquad \alpha \text{ in } F, \quad x \text{ in } \mathfrak{N}.$$

Thus every associator relative to the multiplication in $\mathfrak{J}^+$ is an associator

$$[x_1, x_2, x_3] = (x_1 \cdot x_2) \cdot x_3 - x_1 \cdot (x_2 \cdot x_3), \qquad x_i \text{ in } \mathfrak{N}.$$

We shall first show that $\mathfrak{J}^+$ is associative by showing that the subspace $\mathfrak{P}$ spanned by all of these associators is 0. For any y in $\mathfrak{N}$, (5.44) implies that $(x_1 \cdot x_2)y$ is in $\mathfrak{N}$, so

$$\begin{aligned}
[(x_1 \cdot x_2) \cdot x_3]y = {}& \lambda_3 x_1 \cdot x_2 + (x_1 \cdot x_2) \cdot z_3 \\
& + x_3 \cdot [\lambda_1 x_2 + \lambda_2 x_1 + x_1 \cdot z_2 + x_2 \cdot z_1 \\
& - (x_1 \cdot y) \cdot x_2 - (x_2 \cdot y) \cdot x_1 + (x_1 \cdot x_2) \cdot y] \\
& - [(x_1 \cdot x_2) \cdot y] \cdot x_3 \\
& - (x_3 \cdot y) \cdot (x_1 \cdot x_2) + [(x_1 \cdot x_2) \cdot x_3] \cdot y
\end{aligned}$$

by (5.44). Interchange subscripts 1 and 3, and subtract, to obtain

$$\begin{aligned}
[x_1, x_2, x_3]y = {}& [x_1, x_2, z_3] + [x_1, z_2, x_3] \\
& + [z_1, x_2, x_3] - [x_1 \cdot y, x_2, x_3] - [x_1, x_2 \cdot y, x_3] \\
& + [x_3 \cdot y, x_2, x_1] + [x_1, x_2, x_3] \cdot y
\end{aligned}$$

so that we have the first inclusion in

(5.47) $\mathfrak{P}\mathfrak{N} \subseteq \mathfrak{P} + \mathfrak{P} \cdot \mathfrak{N}, \qquad \mathfrak{N}\mathfrak{P} \subseteq \mathfrak{P} + \mathfrak{P} \cdot \mathfrak{N}.$

The second part of (5.47) follows from $np = -pn + 2p\cdot n$ for p in $\mathfrak{P}$, n in $\mathfrak{N}$.

Define an ascending series $\mathfrak{C}_0 \subseteq \mathfrak{C}_1 \subseteq \mathfrak{C}_2 \subseteq \cdots$ of subspaces $\mathfrak{C}_i$ of $\mathfrak{J}$ by

$$\mathfrak{C}_0 = \mathfrak{P}, \qquad \mathfrak{C}_{i+1} = \mathfrak{C}_i + \mathfrak{C}_i \cdot \mathfrak{N}.$$

Note that all the $\mathfrak{C}_i$ are contained in $\mathfrak{N}$ (actually in $\mathfrak{N} \cdot \mathfrak{N} \cdot \mathfrak{N}$, since $\mathfrak{P}$ is). We prove by induction on i that

(5.48) $\mathfrak{C}_i \mathfrak{N} \subseteq \mathfrak{C}_{i+1}, \qquad \mathfrak{N}\mathfrak{C}_i \subseteq \mathfrak{C}_{i+1}.$

The case $i = 0$ of (5.48) is (5.47). We assume (5.48) and prove that $\mathfrak{C}_{i+1} \mathfrak{N} \subseteq \mathfrak{C}_{i+2}$ as follows: by the assumption of the induction it is sufficient to show

$$(\mathfrak{C}_i \cdot \mathfrak{N})\mathfrak{N} \subseteq \mathfrak{C}_{i+1} + \mathfrak{C}_{i+1} \cdot \mathfrak{N}.$$

Now the flexible law (5.39) is equivalent to

(5.49) $(x \cdot y)z = (yz) \cdot x + (yx) \cdot z - (y \cdot z)x \qquad$ for all $\quad x, y, z$ in $\mathfrak{J}$.

Put x in $\mathfrak{C}_i$, y and z in $\mathfrak{N}$ into (5.49) and use $yz = \mu 1 + w$, μ in F, w in $\mathfrak{N}$, to see that each term of the right-hand side of (5.49) is in $\mathfrak{C}_{i+1} + \mathfrak{C}_{i+1} \cdot \mathfrak{N}$ by the assumption (5.48) of the induction. We have established $\mathfrak{C}_{i+1}\mathfrak{N} \subseteq \mathfrak{C}_{i+2}$. Then, as above,

$$\mathfrak{N}\mathfrak{C}_{i+1} \subseteq \mathfrak{C}_{i+1}\mathfrak{N} + \mathfrak{C}_{i+1} \cdot \mathfrak{N} \subseteq \mathfrak{C}_{i+2},$$

and we have established (5.48). Since $\mathfrak{J}$ is finite-dimensional, there is a positive integer k such that $\mathfrak{C}_{k+1} = \mathfrak{C}_k$. Then $\mathfrak{C}_k$ is an ideal of $\mathfrak{J}$. For $\mathfrak{C}_k \mathfrak{J} = \mathfrak{C}_k(F1 + \mathfrak{N}) \subseteq \mathfrak{C}_k$ by (5.48), and similarly $\mathfrak{J}\mathfrak{C}_k \subseteq \mathfrak{C}_k$. The ideal $\mathfrak{C}_k$, being contained in $\mathfrak{N}$, is not $\mathfrak{J}$. Hence $\mathfrak{C}_k = 0$, since $\mathfrak{J}$ is simple. But $\mathfrak{P} \subseteq \mathfrak{C}_k$, so $\mathfrak{P} = 0$, $\mathfrak{J}^+$ is associative.

An ideal $\mathfrak{B}$ of an algebra $\mathfrak{A}$ is called a *characteristic* ideal (or $\mathfrak{D}$-ideal) in case $\mathfrak{B}$ is mapped into itself by every derivation of $\mathfrak{A}$. $\mathfrak{A}$ is called $\mathfrak{D}$-*simple* if 0 and $\mathfrak{A}$ are the only characteristic ideals of $\mathfrak{A}$.

We show next that the commutative associative algebra $\mathfrak{J}^+$ is $\mathfrak{D}$-simple. Interchange x and y in (5.39) to obtain

(5.50) $(y, x, z) + (z, x, y) = 0 \qquad$ for all $\quad x, y, z$ in $\mathfrak{J}$;

interchange y and z in (5.39) to obtain

(5.51) $(x, z, y) + (y, z, x) = 0$ for all x, y, z in $\mathfrak{J}$;

adding (5.39) and (5.50), and subtracting (5.51), we obtain the identity

(5.52) $[x \cdot y, z] = [x, z] \cdot y + x \cdot [y, z]$ for all x, y, z in $\mathfrak{J}$,

which is valid in any flexible algebra. This identity is equivalent to the statement that

$$D = R_z - L_z \qquad \text{for any} \quad z \text{ in } \mathfrak{J}$$

is a derivation of $\mathfrak{J}^+$. If $\mathfrak{B}$ is an ideal of $\mathfrak{J}^+$, then $x \cdot z$ is in $\mathfrak{B}$ for all x in $\mathfrak{B}$, z in $\mathfrak{J}$. If, furthermore, $\mathfrak{B}$ is characteristic, then $[x, z] = xD$ is in $\mathfrak{B}$, since $D = R_z - L_z$ is a derivation of $\mathfrak{J}^+$. Hence $xz = x \cdot z + \frac{1}{2}[x, z]$ and $zx = x \cdot z - \frac{1}{2}[x, z]$ are in $\mathfrak{B}$ for all x in $\mathfrak{B}$, z in $\mathfrak{J}$; that is, $\mathfrak{B}$ is an ideal of $\mathfrak{J}$. Hence $\mathfrak{J}$ simple implies that the commutative associative algebra $\mathfrak{J}^+$ is $\mathfrak{D}$-simple.

It is a very interesting fact in the theory of commutative associative algebras that, if $\mathfrak{A}$ is a finite-dimensional $\mathfrak{D}$-simple commutative associative algebra of the form $\mathfrak{A} = F1 + \mathfrak{R}$ where $\mathfrak{R}$ is the radical of $\mathfrak{A}$, then (except for the trivial case $\mathfrak{A} = F1$ which may occur at characteristic 0, and which does not give a nodal algebra) F has characteristic p and $\mathfrak{A}$ is the p^n-dimensional algebra $\mathfrak{A} = F[1, x_1, ..., x_n], x_i^p = 0$ (Harper [1]).

Now any derivation D of such an algebra has the form

$$f \to fD = \sum_{i=1}^{n} \frac{\partial f}{\partial x_i} \cdot a_i, \qquad a_i \text{ in } \mathfrak{A},$$

where the a_i of course depend on the derivation D. Now $f \to [f, g]$ is a derivation of $\mathfrak{J}^+$ for any g in $\mathfrak{J}$, so we have

(5.53) $[f, g] = \sum_{i=1}^{n} \dfrac{\partial f}{\partial x_i} \cdot a_i(g), \qquad a_i(g) \text{ in } \mathfrak{J}.$

To evaluate the $a_i(g)$, note that $x_i D = [x_i, g] = a_i(g)$ and

$$[g, x_i] = \sum_{j=1}^{n} \frac{\partial g}{\partial x_j} \cdot a_j(x_i).$$

Then $a_j(x_i) = [x_j, x_i]$ implies

$$a_i(g) = -[g, x_i] = - \sum_{j=1}^{n} \frac{\partial g}{\partial x_j} \cdot [x_j, x_i]$$

or

$$[f, g] = \sum_{i,j=1}^{n} \frac{\partial f}{\partial x_i} \cdot \frac{\partial g}{\partial x_j} \cdot [x_i, x_j]$$

by (5.53), so that $fg = f \cdot g + \frac{1}{2}[f, g]$ implies (5.45) where $c_{ij} = \frac{1}{2}[x_i, x_j]$. If every c_{ij} were in $\mathfrak{N}$, then $\mathfrak{N}$ would be a subalgebra of $\mathfrak{A}$, a contradiction. Hence at least one of the c_{ij} is of the form (5.46) with $\alpha \neq 0$, so it has an inverse, and $n \geq 2$.

Not every algebra described in the conclusion of Theorem 5.6 is simple (Kokoris [5]). However, all such algebras of dimension p^2 are, and for every even n there are simple algebras of dimension p^n. There are relationships between the derivation algebras of nodal noncommutative Jordan algebras and (nonclassical) simple Lie algebras of characteristic p (Schafer [17]). For a general discussion of Lie algebras of characteristic p, see Seligman [1, 3].

Noncommutative Jordan algebras do not in general have Wedderburn decompositions (even in case $\mathfrak{N}^2 = 0$). This may be seen from the example of the 5-dimensional noncommutative Jordan algebra $\mathfrak{J}$ with basis $e_{11}, e_{12}, e_{21}, e_{22}, z$ over F and multiplication table:

	e_{11}	e_{12}	e_{21}	e_{22}	z
e_{11}	e_{11}	e_{12}	z	0	0
e_{12}	0	0	e_{11}	e_{12}	0
e_{21}	$e_{21} - z$	e_{22}	0	z	0
e_{22}	0	0	$e_{21} - z$	e_{22}	z
z	z	0	0	0	0

The radical $\mathfrak{N}$ of $\mathfrak{J}$ is $\mathfrak{N} = Fz$. Also $\mathfrak{N}^2 = 0$, and $\mathfrak{J}/\mathfrak{N} \cong F_2$ with basis $\bar{e}_{11}, \bar{e}_{12}, \bar{e}_{21}, \bar{e}_{22}$. Suppose there were a subalgebra $\mathfrak{S} \cong \mathfrak{J}/\mathfrak{N}$. Then $\mathfrak{S}$ would have a usual matrix basis $g_{11}, g_{12}, g_{21}, g_{22}$, and there would be an automorphism H of $\mathfrak{J}/\mathfrak{N}$ such that $\bar{g}_{ij} = \bar{e}_{ij} H$ $(i, j = 1, 2)$. But any automorphism of $\mathfrak{J}/\mathfrak{N}$ is inner (Albert [24], p. 51). Thus there is an invertible element $\bar{y} = \alpha\bar{e}_{11} + \beta\bar{e}_{12} + \gamma\bar{e}_{21} + \delta\bar{e}_{22}$ $(\Delta = \alpha\delta - \beta\gamma \neq 0)$ in $\mathfrak{J}/\mathfrak{N}$ such that $\bar{g}_{ij} = \bar{y}\,\bar{e}_{ij}\,\bar{y}^{-1}$. It follows that

$$g_{11} = \Delta^{-1}(\alpha\delta e_{11} - \alpha\beta e_{12} + \gamma\delta e_{21} - \beta\gamma e_{22}) + \varepsilon z,$$

$$g_{12} = \Delta^{-1}(-\alpha\gamma e_{11} + \alpha^2 e_{12} - \gamma^2 e_{21} + \alpha\gamma e_{22}) + \kappa z,$$

$$g_{21} = \Delta^{-1}(\beta\delta e_{11} - \beta^2 e_{12} + \delta^2 e_{21} - \beta\delta e_{22}) + \lambda z,$$

$$g_{22} = \Delta^{-1}(-\beta\gamma e_{11} + \alpha\beta e_{12} - \gamma\delta e_{21} + \alpha\delta e_{22}) + \mu z.$$

Equating coefficients of z in $g_{11}\,g_{12} = g_{12}$, $g_{22}\,g_{12} = g_{11}\,g_{21} = 0$, $g_{22}\,g_{21} = g_{21}$, $g_{21}\,g_{12} = g_{22}$ yields equations in $\alpha, \beta, ..., \mu$ which force $\Delta = 0$, a contradiction.

Noncommutative Jordan algebras are a natural generalization of both types of algebras with which we have become most familiar: alternative and Jordan algebras. In the Introduction we indicated that numerous generalizations of familiar classes of algebras have been made. As the reader who consults some of the papers cited in the Bibliography will discover, not all of these algebras are power-associative. However, we conclude this chapter (and our book) with brief mention of another class of power-associative algebras.

An algebra $\mathfrak{A}$ over F is called *right alternative* in case the right alternative law

$$yx^2 = (yx)x \qquad \text{for all} \quad x, y \text{ in } \mathfrak{A}$$

is satisfied. It is easy to see that any right alternative algebra $\mathfrak{A}$ of characteristic $\neq 2$ is power-associative. Also it follows from the first part of (3.69) that $\mathfrak{A}^+$ is a special Jordan algebra. By a modification of the technique used in Theorem 5.4, Albert has shown that the radical $\mathfrak{N}$ (= maximal nilideal) of any finite-dimensional right alternative algebra $\mathfrak{A}$ of characteristic $\neq 2$ coincides with the radical of $\mathfrak{A}^+$, that $\mathfrak{A}/\mathfrak{N}$ is a direct sum of simple ideals, and that every semisimple right alternative algebra is actually alternative (Albert [9, 18]).

BIBLIOGRAPHY

Albert, A. A.

[1] On a certain algebra of quantum mechanics, *Ann. of Math.* **35**, 65–73 (1934).

[2] Quadratic forms permitting composition, *Ann. of Math.* **43**, 161–177 (1942).

[3] Non-associative algebras, I, *Ann. of Math.* **43**, 685–707 (1942); II, *ibid.* 708–723.

[4] The radical of a non-associative algebra, *Bull. Amer. Math. Soc.* **48**, 891–897 (1942).

[5] On Jordan algebras of linear transformations, *Trans. Amer. Math. Soc.* **59**, 524–555 (1946).

[6] A structure theory for Jordan algebras, *Ann. of Math.* **48**, 546–567 (1947).

[7] On the power-associativity of rings, *Summa Brasil. Math.* **2**, 21–32 (1948).

[8] Power-associative rings, *Trans. Amer. Math. Soc.* **64**, 552–593 (1948).

[9] On right alternative algebras, *Ann. of Math.* **50**, 318–328 (1949).

[10] A theory of trace-admissible algebras, *Proc. Nat. Acad. Sci. U.S.A.* **35**, 317–322 (1949).

[11] Absolute-valued algebraic algebras, *Bull. Amer. Math. Soc.* **55**, 763–768 (1949); A note of correction, *Bull. Amer. Math. Soc.* **55**, 1191 (1949).

[12] A note on the exceptional Jordan algebra, *Proc. Nat. Acad. Sci. U.S.A.* **36**, 372–374 (1950).

[13] A theory of power-associative commutative algebras, *Trans. Amer. Math. Soc.* **69**, 503–527 (1950).

[14] New simple power-associative algebras, *Summa Brasil. Math.* **2**, 183–194 (1951).

[15] On simple alternative rings, *Canad. J. Math.* **4**, 129–135 (1952).

[16] On nonassociative division algebras, *Trans. Amer. Math. Soc.* **72**, 296–309 (1952).

[17] On commutative power-associative algebras of degree two, *Trans. Amer. Math. Soc.* **74**, 323–343 (1953).

[18] The structure of right alternative algebras, *Ann. of Math.* (2) **59**, 408–417 (1954).

[19] On partially stable algebras, *Trans. Amer. Math. Soc.* **84**, 430–443 (1957); Addendum to the paper on partially stable algebras, *ibid*, **87**, 57–62 (1958).

[20] A construction of exceptional Jordan division algebras, *Ann. of Math.* (2) **67**, 1–28 (1958).

[21] Finite noncommutative division algebras, *Proc. Amer. Math. Soc.* **9**, 928–932 (1958).

[22] A solvable exceptional Jordan algebra, *J. Math. Mech.* **8**, 331–337 (1959).

[23] Generalized twisted fields, *Pacific J. Math.* **11**, 1–8 (1961).

[24] "Structure of Algebras" (*Colloq. Publ.*, Vol. 24) *Amer. Math. Soc.*, Providence, 1939.

Albert, A. A., and Jacobson, N.

[1] On reduced exceptional simple Jordan algebras, *Ann. of Math.* (2) **66**, 400–417 (1957).

Albert, A. A., and Paige, L. J.

[1] On a homomorphism property of certain Jordan algebras, *Trans. Amer. Math. Soc.* **93**, 20–29 (1959).

Artin, E.

[1] "Galois Theory" (Notre Dame Mathematical Lectures, No. 2) 2nd ed., Notre Dame, 1948.

[2] "Geometric Algebra," Wiley (Interscience), New York, 1957.

Askinuze, V. G.

[1] A theorem on the splittability of J-algebras, *Ukrain. Mat. Z.* (*Russian*) **3**, 381–398 (1951).

Barnes, R. T.

[1] On derivation algebras and Lie algebras of prime characteristic, Yale dissertation, 1963.

Birkhoff, G., and Whitman, P. M.

[1] Representations of Jordan and Lie algebras, *Trans. Amer. Math. Soc.* **65**, 116–136 (1949).

van der Blij, F.

[1] History of the octaves, *Simon Stevin* **34**, 106–125 (1961).

van der Blij, F., and Springer, T. A.

[1] The arithmetics of octaves and of the group G_2, *Nederl. Akad. Wetensch. Proc. Ser. A.* **62**, 406–418 (1959).

[2] Octaves and triality, *Nieuw Arch. Wisk.* **8**, 158–169 (1960).

Bott, R., and Milnor, J.

[1]   On the parallelizability of the spheres, *Bull. Amer. Math. Soc.* **64**, 87–89 (1958)

Braun, H., and Koecher, M.

[1]   "Jordan-Algebren," Springer-Verlag, 1966.

Brown, B., and McCoy, N. H.

[1]   Prime ideals in nonassociative rings, *Trans. Amer. Math. Soc.* **89**, 245–255 (1958).

Brown, R. B.

[1]   A new type of nonassociative algebras, *Proc. Nat. Acad. Sci. U.S.A.* **50**, 947–949 (1963).

Bruck, R. H.

[1]   Some results in the theory of linear nonassociative algebras, *Trans. Amer. Math. Soc.* **56**, 141–199 (1944).
[2]   Recent advances in the foundations of euclidean plane geometry, *Amer. Math. Monthly* **62**, No. 7, part II, 2–17 (1955).

Bruck, R. H., and Kleinfeld, E.

[1]   The structure of alternative division rings, *Proc. Amer. Math. Soc.* **2**, 878–890 (1951).

Chevalley, C., and Schafer, R. D.

[1]   The exceptional simple Lie algebras F_4 and E_6, *Proc. Nat. Acad. Sci. U.S.A.* **36**, 137–141 (1950).

Cohn, P. M.

[1]   On homomorphic images of special Jordan algebras, *Canad. J. Math.* **6**, 253–264 (1954).
[2]   Two embedding theorems for Jordan algebras, *Proc. London Math. Soc.* (3) **9**, 503–524 (1959).
[3]   "Universal Algebra." Harper and Row, New York, 1965.

Dickson, L. E.

[1]   On quaternions and their generalization and the history of the eight square theorem, *Ann. of Math.* **20**, 155–171 (1919).

Eilenberg, S.

[1]   Extensions of general algebras, *Ann. Soc. Polon. Math.* **21**, 125–134 (1948).

Etherington, I. M. H.

[1] Genetic algebras, *Proc. Roy. Soc. Edinburgh* **59**, 242–258 (1939).
[2] Special train algebras, *Quart. J. Math. Oxford. Ser.* **12**, 1–8 (1941).
[3] Nonassociative algebra and the symbolism of genetics, *Proc. Roy. Soc. Edinburgh* **61**, 24–42 (1941).

Freudenthal, H.

[1] Oktaven, Ausnahmengruppen und Oktavengeometrie, Utrecht, 1951.
[2] Zur ebenen Oktavengeometrie, *Nederl. Akad. Wetensch. Proc. Ser. A.* **56**, 195–200 (1953).
[3] Beziehungen der E_7 und E_8 zur Octavenebene, I, *Nederl. Akad. Wetensch. Proc. Ser. A.* **57**, 218–230 (1954); II, *ibid.* 363–368; III, *ibid.* **58**, 151–157 (1955); IV, *ibid.* 277–285; V, *ibid.* **62**, 165–179 (1959); VI, *ibid.* 180–191; VII, *ibid.* 192–201; VIII, *ibid.* 447–465; IX, *ibid.* 466–474; X, *ibid.* **66**, 457–471 (1963); XI, *ibid.* 472–487.
[4] Lie groups in the foundations of geometry, *Advan. Math.* **1**, 145–190 (1964).

Gerstenhaber, M.

[1] On nilalgebras and linear varieties of nilpotent matrices, II, *Duke Math. J.* **27**, 21–32 (1960).

Glennie, C. M.

[1] Identities in Jordan algebras, Yale dissertation, 1963.

Hall, M., Jr.

[1] Projective planes, *Trans. Amer. Math. Soc.* **54**, 229–277 (1943); Correction, *ibid.* **65**, 474 (1949).
[2] Projective planes and related topics, Calif. Inst. of Tech., 1954.

Harper, L. R., Jr.

[1] On differentiably simple algebras, *Trans. Amer. Math. Soc.* **100**, 63–72 (1961).

Harris, B.

[1] Centralizers in Jordan algebras, *Pacific J. Math.* **8**, 757–790 (1958).
[2] Derivations of Jordan algebras, *Pacific J. Math.* **9**, 495–512 (1959).

Herstein, I. N.

[1] Lie and Jordan structures in simple, associative rings, *Bull. Amer. Math. Soc.* **67**, 517–531 (1961).

Hirzebruch, U.
[1] Über Jordan-Algebren und Riemannsche symmetrische Räume vom Rang 1, *Math. Z.* **90**, 339–354 (1965).

Hochschild, G. P.

[1] Semisimple algebras and generalized derivations, *Amer. J. Math.* **64**, 677–694 (1942).

Hurwitz, A.

[1] Über die Komposition der quadratischen Formen von beliebig vielen Variabeln, *Nachr. Ges. Wiss. Göttingen*, pp. 309–316 (1898).

Jacobson, F. D., and Jacobson, N.

[1] Classification and representation of semisimple Jordan algebras, *Trans. Amer. Math. Soc.* **65**, 141–169 (1949).

Jacobson, N.

[1] A note on nonassociative algebras, *Duke Math. J.* **3**, 544–548 (1937).

[2] Abstract derivation and Lie algebras, *Trans. Amer. Math. Soc.* **42**, 206–224 (1937).

[3] Cayley numbers and normal simple Lie algebras of type *G*, *Duke Math. J.* **5**, 775–783 (1939).

[4] Structure theory of simple rings without finiteness assumptions, *Trans. Amer. Math. Soc.* **57**, 228–245 (1945).

[5] Isomorphisms of Jordan rings, *Amer. J. Math.* **70**, 317–326 (1948).

[6] The center of a Jordan ring, *Bull. Amer. Math. Soc.* **54**, 316–322 (1948).

[7] Derivation algebras and multiplication algebras of semisimple Jordan algebras, *Ann. of Math.* **50**, 866–874 (1949).

[8] Lie and Jordan triple systems, *Amer. J. Math.* **71**, 149–170 (1949).

[9] General representation theory of Jordan algebras, *Trans. Amer. Math. Soc.* **70**, 509–530 (1951).

[10] Operator commutativity in Jordan algebras, *Proc. Amer. Math. Soc.* **3**, 973–976 (1952).

[11] Structure of alternative and Jordan bimodules, *Osaka Math. J.* **6**, 1–71 (1954).

[12] A Kronecker factorization theorem for Cayley algebras and the exceptional simple Jordan algebra, *Amer. J. Math.* **76**, 447–452 (1954).

[13] A theorem on the structure of Jordan algebras, *Proc. Nat. Acad. Sci. U.S.A.* **42**, 140–147 (1956).

[14] Composition algebras and their automorphisms, *Rend. Circ. Mat. Palermo* (2) **7**, 55–80 (1958).

[15] Nilpotent elements in semisimple Jordan algebras, *Math. Ann.* **136**, 375–386 (1958).

[16] Some groups of transformations defined by Jordan algebras, I, *J. reine angew. Math.* **201**, 178–195 (1959); II, *ibid.* **204**, 74–98 (1960); III. *ibid.* **207**, 61–85 (1961).

[17] Exceptional Lie algebras, dittoed, 57 pp.

[18] Cayley planes, dittoed, 28 pp.
[19] A coordinatization theorem for Jordan algebras, *Proc. Nat. Acad. Sci. U.S.A.* **48**, 1154–1160 (1962).
[20] MacDonald's theorem on Jordan algebras, *Arch. Math.* **13**, 241–250 (1962).
[21] Generic norm of an algebra, *Osaka Math. J.* **15**, 25–50 (1963).
[22] Triality and Lie algebras of type D_4, *Rend. Circ. Mat. Palermo* (2) **13**, 129-153 (1964).
[23] Structure theory for a class of Jordan algebras, *Proc. Nat. Acad. Sci. U.S.A.* **55**, 243–251 (1966).
[24] "Lectures in Abstract Algebra," Vol. II. Van Nostrand, Princeton, 1953.
[25] "Lie Algebras," Wiley (Interscience), New York, 1962.

Jacobson, N., and Paige, L. J.

[1] On Jordan algebras with two generators, *J. Math. Mech.* **6**, 895–906 (1957).

Jordan, P.

[1] Über eine Klasse nichtassoziativer hyperkomplexer Algebren, *Nachr. Ges. Wiss. Göttingen*, pp. 569–575 (1932).
[2] Über Verallgemeinerungsmöglichkeiten des Formalismus der Quantenmechanik, *Nachr. Ges. Wiss. Göttingen*, pp. 209–214 (1933).
[3] Über die Multiplication quantenmechanischer Grössen, *Z. Physik* **80**, 285–291 (1933).
[4] Über eine nicht-desarguessche ebene projecktive Geometrie, *Abh. Math. Sem. Univ. Hamburg* **16**, 74–76 (1949).

Jordan, P., von Neumann, J., and Wigner, E.

[1] On an algebraic generalization of the quantum mechanical formalism, *Ann. of Math.* **35**, 29–64 (1934).

Kalisch, G. K.

[1] On special Jordan algebras, *Trans. Amer. Math. Soc.* **61**, 482–494 (1947).

Kaplansky, I.

[1] Infinite-dimensional quadratic forms permitting composition, *Proc. Amer. Math. Soc.* **4**, 956–960 (1953).

Kleinfeld, E.

[1] Right alternative rings, *Proc. Amer. Math. Soc.* **4**, 939–944 (1953).
[2] Simple alternative rings, *Ann. of Math.* **58**, 544–547 (1953).
[3] Primitive alternative rings and semisimplicity, *Amer. J. Math.* **77**, 725–730 (1955).
[4] Standard and accessible rings, *Canad. J. Math.* **8**, 335–340 (1956).

[5] Alternative nil rings, *Ann. of Math.* (2) **66**, 395–399 (1957).

[6] On alternative and right alternative rings, Report of a Conference on Linear Algebras, *Nat. Acad. Sci.–Nat. Res. Council, Publ. 502*, pp. 20–23 (1957).

[7] Quasi-nil rings, *Proc. Amer. Math. Soc.* **10**, 477–479 (1959).

[8] Simple algebras of type (1, 1) are associative, *Canad. J. Math.* **13**, 129–148 (1961).

[9] A characterization of the Cayley numbers, *M.A.A. Studies in Math.* **2**, 126–143 (1963).

[10] Middle nucleus-center in a simple Jordan ring, *J. Algebra* **1**, 40–42 (1964).

Kleinfeld, E., and Kokoris, L. A.

[1] Flexible algebras of degree one, *Proc. Amer. Math. Soc.* **13**, 891–893 (1962).

Kleinfeld, E., Kosier, F., Osborn, J. M., and Rodabaugh, D.

[1] The structure of associator dependent rings, *Trans. Amer. Math. Soc.* **110**, 473–483 (1964).

Koecher, M.

[1] Jordan algebras and their applications (dittoed), University of Minnesota, 1962.

[2] Eine Characterisierung der Jordan Algebren, *Math. Ann.* **148**, 244–256 (1962).

Kokoris, L. A.

[1] Power-associative commutative algebras of degree two, *Proc. Nat. Acad. Sci. U.S.A.* **38**, 534–537 (1952).

[2] New results on power-associative algebras, *Trans. Amer. Math. Soc.* **77**, 363–373 (1954).

[3] Simple power-associative algebras of degree two, *Ann. of Math.* (2) **64**, 544–550 (1956).

[4] Simple nodal noncommutative Jordan algebras, *Proc. Amer. Math. Soc.* **9**, 652–654 (1958).

[5] Nodal noncommutative Jordan algebras, *Canad. J. Math.* **12**, 488–492 (1960).

[6] Flexible nilstable algebras, *Proc. Amer. Math. Soc.* **13**, 335–340 (1962).

Kosier, F.

[1] On a class of nonflexible algebras, *Trans. Amer. Math. Soc.* **102**, 299–318 (1962).

Kosier, F., and Osborn, J. M.

[1] Nonassociative algebras satisfying identities of degree three, *Trans. Amer. Math. Soc.* **110**, 484–492 (1964).

Kurosch, A.

[1] The present status of the theory of rings and algebras, *Uspehi Mat. Nauk* (*N.S.*) (*Russian*) **6**, no. 2 (42), 3–15 (1951).

Laufer, P. J., and Tomber, M. L.

[1] Some Lie admissible algebras, *Canad. J. Math.* **14**, 287–292 (1962).

Leadley, J. D., and Ritchie, R. W.

[1] Conditions for the power-associativity of algebras, *Proc. Amer. Math. Soc.* **11**, 399–405 (1960).

Lister, W. G.

[1] A structure theory of Lie triple systems, *Trans. Amer. Math. Soc.* **72**, 217–242 (1952).

Macdonald, I. G.

[1] Jordan algebras with three generators, *Proc. London Math. Soc.* (3) **10**, 395–408 (1960).

Malcev, A.

[1] On the representation of an algebra as a direct sum of the radical and a semi-simple subalgebra, *C. R.* (*Doklady*) *Acad. Sci. URSS* **36**, 42–45 (1942).

Maneri, C.

[1] Simple $(-1, 1)$ rings with an idempotent, *Proc. Amer. Math. Soc.* **14**, 110–117 (1963).

McCrimmon, K.

[1] Jordan algebras of degree 1, *Bull. Amer. Math. Soc.* **70**, 702 (1964).
[2] Norms and noncommutative Jordan algebras, *Pacific J. Math.* **15**, 925–956 (1965).
[3] Structure and representations of noncommutative Jordan algebras, *Trans. Amer. Math. Soc.*, **121**, 187–199 (1966).
[4] A proof of Schafer's conjecture for infinite-dimensional forms admitting composition, *J. Algebra.*, to appear.
[5] Finite power-associative division rings, *Proc. Amer. Math. Soc.*, to appear.

Meyberg, K.

[1] Über die Spur der quadratischen Darstellung von Jordan–Algebren, Munich dissertation, 1964.

Mills, W. H.

[1] A theorem on the representation theory of Jordan algebras, *Pacific J. Math.* **1**, 255–264 (1951).

Moufang, R.

[1] Alternativkörper und der Satz vom vollständigen Vierseit (D_9), *Abh. Math. Sem. Univ. Hamburg.* **9**, 207–222 (1933).
[2] Zur Struktur von Alternativkörpern, *Math. Ann.* **110**, 416–430 (1935).

von Neumann, J.

[1] On an algebraic generalization of the quantum mechanical formalism, *Mat. Sb.* **1**, 415–482 (1936).

Oehmke, R. H.

[1] On flexible algebras, *Ann. of Math.* (2) **68**, 221–230 (1958).
[2] On flexible power-associative algebras of degree two, *Proc. Amer. Math. Soc.* **12**, 151–158 (1961).
[3] On commutative algebras of degree two, *Trans. Amer. Math. Soc.* **105**, 295–313 (1962).
[4] Nodal noncommutative Jordan algebras, *Trans. Amer. Math. Soc.* **112**, 416–431 (1964).

Oehmke, R. H., and Sandler, R.

[1] The collineation groups of division ring planes, *J. reine angew. Math.* **216**, 67–87 (1964).

Osborn, J. M.

[1] Quadratic division algebras, *Trans. Amer. Math. Soc.* **105**, 202–221 (1962).
[2] Identities of nonassociative algebras, *Canad. J. Math.* **17**, 78–92 (1965).

Paige, L. J.

[1] Jordan algebras, *M.A.A. Studies in Math.* **2**, 144–186 (1963).

Penico, A. J.

[1] The Wedderburn principal theorem for Jordan algebras, *Trans. Amer. Math. Soc.* **70**, 404–420 (1951).

Sagle, A. A.

[1] Malcev algebras, *Trans. Amer. Math. Soc.* **101**, 426–458 (1961).

[2]   Simple Malcev algebras over fields of characteristic 0, *Pacific J. Math.* **12**, 1057–1078 (1962).

Schafer, R. D.

[1]   Alternative algebras over an arbitrary field, *Bull. Amer. Math. Soc.* **49**, 549–555 (1943).

[2]   Concerning automorphisms of nonassociative algebras, *Bull. Amer. Math. Soc.* **53**, 573–583 (1947).

[3]   The exceptional simple Jordan algebras, *Amer. J. Math.* **70**, 82–94 (1948).

[4]   Structure of genetic algebras, *Amer. J. Math.* **71**, 121–135 (1949).

[5]   The Wedderburn principal theorem for alternative algebras, *Bull. Amer. Math. Soc.* **55**, 604–614 (1949).

[6]   Inner derivations of nonassociative algebras, *Bull. Amer. Math. Soc.* **55**, 769–776 (1949).

[7]   A theorem on the derivations of Jordan algebras, *Proc. Amer. Math. Soc.* **2**, 290–294 (1951).

[8]   Representations of alternative algebras, *Trans. Amer. Math. Soc.* **72**, 1–17 (1952).

[9]   The Casimir operation for alternative algebras, *Proc. Amer. Math. Soc.* **4**, 444–451 (1953).

[10]   A generalization of a theorem of Albert, *Proc. Amer. Math. Soc.* **4**, 452–455 (1953).

[11]   On the algebras formed by the Cayley–Dickson process, *Amer. J. Math.* **76**, 435–446 (1954).

[12]   Noncommutative Jordan algebras of characteristic 0, *Proc. Amer. Math. Soc.* **6**, 472–475 (1955).

[13]   Structure and representation of nonassociative algebras, *Bull. Amer. Math. Soc.* **61**, 469–484 (1955).

[14]   On noncommutative Jordan algebras, *Proc. Amer. Math. Soc.* **9**, 110–117 (1958).

[15]   Restricted noncommutative Jordan algebras of characteristic p, *Proc. Amer. Math. Soc.* **9**, 141–144 (1958).

[16]   On cubic forms permitting composition, *Proc. Amer. Math. Soc.* **10**, 917–925 (1959).

[17]   Nodal noncommutative Jordan algebras and simple Lie algebras of characteristic p, *Trans. Amer. Math. Soc.* **94**, 310–326 (1960).

[18]   Cubic forms permitting a new type of composition, *J. Math. Mech.* **10**, 159–174 (1961).

[19]   On forms of degree n permitting composition, *J. Math. Mech.* **12**, 777–792 (1963).

[20]   On the simplicity of the Lie algebras E_7 and E_8, *Nederl. Akad. Wetensch. Proc. Ser. A.* **69**, 64–69 (1966).

Seligman, G. B.

[1]   A survey of Lie algebras of characteristic p, Report of a Conference on Linear Algebras, *Nat. Acad. Sci.–Nat. Res. Council, Publ. 502* pp. 24–32 (1957).

[2] On automorphisms of Lie algebras of classical type, *Trans. Amer. Math. Soc.* **92**, 430–448 (1959); II, *ibid.* **94**, 452–482 (1960); III, *ibid.* **97**, 286–316 (1960).

[3] "Modular Lie Algebras", Springer–Verlag, to appear in 1967.

Shirshov, A. I.

[1] On special *J*-rings, *Mat. Sb.* (*Russian*) **38** (80) 149–166 (1956).

[2] Some questions in the theory of rings close to associative, *Uspehi Mat. Nauk* (*Russian*) **13**, no. 6 (84), 3–20 (1958).

Skornyakov, L. A.

[1] Alternative fields, *Ukrain. Mat. Ž.* (*Russian*) **2**, 70–85 (1950).

[2] Right-alternative fields, *Izv. Akad. Nauk SSSR Ser. Mat.* (*Russian*) **15**, 177–184 (1951).

Smiley, M. F.

[1] The radical of an alternative ring, *Ann. of Math.* **49**, 702–709 (1948).

[2] Application of a radical of Brown and McCoy to nonassociative rings, *Amer. J. Math.* **72**, 93–100 (1950).

[3] Some questions concerning alternative rings, *Bull. Amer. Math. Soc.* **57**, 36–43 (1951).

[4] Jordan homomorphisms and right alternative rings, *Proc. Amer. Math. Soc.* **8**, 668–671 (1957).

Springer, T. A.

[1] On a class of Jordan algebras, *Nederl. Akad. Wetensch. Proc. Ser. A.* **62**, 254–264 (1959).

[2] The projective octave plane, *Nederl. Akad. Wetensch. Proc. Ser. A.* **63**, 74–101 (1960).

[3] The classification of reduced exceptional simple Jordan algebras, *Nederl. Akad. Wetensch. Proc. Ser. A.* **63**, 414–422 (1960).

[4] Characterization of a class of cubic forms, *Nederl. Akad. Wetensch. Proc. Ser. A.* **65**, 259–265 (1962).

[5] On the geometric algebra of the octave planes, *Nederl. Akad. Wetensch. Proc. Ser. A.* **65**, 451–468 (1962).

Springer, T. A., and Veldkamp, F. D.

[1] Elliptic and hyperbolic octave planes, *Nederl. Akad. Wetensch. Proc. Ser. A.* **66**, 413–451 (1963).

Taft, E. J.

[1] Invariant Wedderburn factors, *Illinois J. Math.* **1**, 565–573 (1957).

[2]   The Whitehead first lemma for alternative algebras, *Proc. Amer. Math. Soc.* **8**, 950–956 (1957).

Tits, J.

[1]   Sur la trialité et les algèbres d'octaves, *Acad. Roy. Belg. Bull. Cl. Sci.* (5) **44**, 332–350 (1958).
[2]   Une classe d'algèbres de Lie en relation avec les algèbres de Jordan, *Nederl. Akad. Wetensch. Proc. Ser. A* **65**, 530–535 (1962).
[3]   A theorem on generic norms of strictly power associative algebras, *Proc. Amer. Math. Soc.* **15**, 35–36 (1964).
[4]   Algèbres alternatives, algèbres de Jordan, et algèbres de Lie exceptionnelles, summary of a manuscript to appear.

Tomber, M. L.

[1]   Lie algebras of type *F*, *Proc. Amer. Math. Soc.* **4**, 759–768 (1953).

Topping, D. M.
[1]   Jordan algebras of self-adjoint operators, *Mem. Amer. Math. Soc. No. 53*, 1965.

Urbanik, K., and Wright, F. B.

[1]   Absolute-valued algebras, *Proc. Amer. Math. Soc.* **11**, 861–866 (1960).

Vinberg, E. B.

[1]   Homogeneous cones, *Dokl. Akad. Nauk. SSSR.* **133**, 9–12 (1960).
[2]   The theory of homogeneous convex cones, *Trudy Moskov. Mat. Obšč.* (*Russian*) **12**, 303–358 (1963).

Zariski, O., and Samuel, P.
[1]   "Commutative Algebra," Vol. I. Van Nostrand, Princeton, 1958.

Zorn, M.

[1]   Theorie der alternativen Ringe, *Abh. Math. Sem. Univ. Hamburg.* **8**, 123–147 (1930).
[2]   Alternativkörper und quadratische systeme, *Abh. Math. Sem. Univ. Hamburg.* **9**, 395–402 (1933).
[3]   The automorphisms of Cayley's nonassociative algebra, *Proc. Nat. Acad. Sci. U.S.A.* **21**, 355–358 (1935).
[4]   Alternative rings and related questions I: Existence of the radical, *Ann. of Math.* **42**, 676–686 (1941).

Index

A CATALOG OF SELECTED
DOVER BOOKS
IN SCIENCE AND MATHEMATICS

A CATALOG OF SELECTED
DOVER BOOKS
IN SCIENCE AND MATHEMATICS

QUALITATIVE THEORY OF DIFFERENTIAL EQUATIONS, V.V. Nemytskii and V.V. Stepanov. Classic graduate-level text by two prominent Soviet mathematicians covers classical differential equations as well as topological dynamics and ergodic theory. Bibliographies. 523pp. 5⅜ × 8½. 65954-2 Pa. $10.95

MATRICES AND LINEAR ALGEBRA, Hans Schneider and George Phillip Barker. Basic textbook covers theory of matrices and its applications to systems of linear equations and related topics such as determinants, eigenvalues and differential equations. Numerous exercises. 432pp. 5⅜ × 8½. 66014-1 Pa. $10.95

QUANTUM THEORY, David Bohm. This advanced undergraduate-level text presents the quantum theory in terms of qualitative and imaginative concepts, followed by specific applications worked out in mathematical detail. Preface. Index. 655pp. 5⅜ × 8½. 65969-0 Pa. $13.95

ATOMIC PHYSICS (8th edition), Max Born. Nobel laureate's lucid treatment of kinetic theory of gases, elementary particles, nuclear atom, wave-corpuscles, atomic structure and spectral lines, much more. Over 40 appendices, bibliography. 495pp. 5⅜ × 8½. 65984-4 Pa. $12.95

ELECTRONIC STRUCTURE AND THE PROPERTIES OF SOLIDS: The Physics of the Chemical Bond, Walter A. Harrison. Innovative text offers basic understanding of the electronic structure of covalent and ionic solids, simple metals, transition metals and their compounds. Problems. 1980 edition. 582pp. 6⅛ × 9¼. 66021-4 Pa. $15.95

BOUNDARY VALUE PROBLEMS OF HEAT CONDUCTION, M. Necati Özisik. Systematic, comprehensive treatment of modern mathematical methods of solving problems in heat conduction and diffusion. Numerous examples and problems. Selected references. Appendices. 505pp. 5⅜ × 8½. 65990-9 Pa. $12.95

A SHORT HISTORY OF CHEMISTRY (3rd edition), J.R. Partington. Classic exposition explores origins of chemistry, alchemy, early medical chemistry, nature of atmosphere, theory of valency, laws and structure of atomic theory, much more. 428pp. 5⅜ × 8½. (Available in U.S. only) 65977-1 Pa. $10.95

A HISTORY OF ASTRONOMY, A. Pannekoek. Well-balanced, carefully reasoned study covers such topics as Ptolemaic theory, work of Copernicus, Kepler, Newton, Eddington's work on stars, much more. Illustrated. References. 521pp. 5⅜ × 8½. 65994-1 Pa. $12.95

PRINCIPLES OF METEOROLOGICAL ANALYSIS, Walter J. Saucier. Highly respected, abundantly illustrated classic reviews atmospheric variables, hydrostatics, static stability, various analyses (scalar, cross-section, isobaric, isentropic, more). For intermediate meteorology students. 454pp. 6⅛ × 9¼. 65979-8 Pa. $14.95

RELATIVITY, THERMODYNAMICS AND COSMOLOGY, Richard C. Tolman. Landmark study extends thermodynamics to special, general relativity; also applications of relativistic mechanics, thermodynamics to cosmological models. 501pp. 5⅜ × 8½.　65383-8 Pa. $12.95

APPLIED ANALYSIS, Cornelius Lanczos. Classic work on analysis and design of finite processes for approximating solution of analytical problems. Algebraic equations, matrices, harmonic analysis, quadrature methods, much more. 559pp. 5⅜ × 8½.　65656-X Pa. $13.95

SPECIAL RELATIVITY FOR PHYSICISTS, G. Stephenson and C.W. Kilmister. Concise elegant account for nonspecialists. Lorentz transformation, optical and dynamical applications, more. Bibliography. 108pp. 5⅜ × 8½.　65519-9 Pa. $4.95

INTRODUCTION TO ANALYSIS, Maxwell Rosenlicht. Unusually clear, accessible coverage of set theory, real number system, metric spaces, continuous functions, Riemann integration, multiple integrals, more. Wide range of problems. Undergraduate level. Bibliography. 254pp. 5⅜ × 8½.　65038-3 Pa. $7.95

INTRODUCTION TO QUANTUM MECHANICS With Applications to Chemistry, Linus Pauling & E. Bright Wilson, Jr. Classic undergraduate text by Nobel Prize winner applies quantum mechanics to chemical and physical problems. Numerous tables and figures enhance the text. Chapter bibliographies. Appendices. Index. 468pp. 5⅜ × 8½.　64871-0 Pa. $11.95

ASYMPTOTIC EXPANSIONS OF INTEGRALS, Norman Bleistein & Richard A. Handelsman. Best introduction to important field with applications in a variety of scientific disciplines. New preface. Problems. Diagrams. Tables. Bibliography. Index. 448pp. 5⅜ × 8½.　65082-0 Pa. $12.95

MATHEMATICS APPLIED TO CONTINUUM MECHANICS, Lee A. Segel. Analyzes models of fluid flow and solid deformation. For upper-level math, science and engineering students. 608pp. 5⅜ × 8½.　65369-2 Pa. $13.95

ELEMENTS OF REAL ANALYSIS, David A. Sprecher. Classic text covers fundamental concepts, real number system, point sets, functions of a real variable, Fourier series, much more. Over 500 exercises. 352pp. 5⅜ × 8½. 65385-4 Pa. $10.95

PHYSICAL PRINCIPLES OF THE QUANTUM THEORY, Werner Heisenberg. Nóbel Laureate discusses quantum theory, uncertainty, wave mechanics, work of Dirac, Schroedinger, Compton, Wilson, Einstein, etc. 184pp. 5⅜ × 8½. 60113-7 Pa. $5.95

INTRODUCTORY REAL ANALYSIS, A.N. Kolmogorov, S.V. Fomin. Translated by Richard A. Silverman. Self-contained, evenly paced introduction to real and functional analysis. Some 350 problems. 403pp. 5⅜ × 8½.　61226-0 Pa. $9.95

PROBLEMS AND SOLUTIONS IN QUANTUM CHEMISTRY AND PHYSICS, Charles S. Johnson, Jr. and Lee G. Pedersen. Unusually varied problems, detailed solutions in coverage of quantum mechanics, wave mechanics, angular momentum, molecular spectroscopy, scattering theory, more. 280 problems plus 139 supplementary exercises. 430pp. 6½ × 9¼.　65236-X Pa. $12.95

CATALOG OF DOVER BOOKS

ASYMPTOTIC METHODS IN ANALYSIS, N.G. de Bruijn. An inexpensive, comprehensive guide to asymptotic methods—the pioneering work that teaches by explaining worked examples in detail. Index. 224pp. 5⅜ × 8½.   64221-6 Pa. $6.95

OPTICAL RESONANCE AND TWO-LEVEL ATOMS, L. Allen and J.H. Eberly. Clear, comprehensive introduction to basic principles behind all quantum optical resonance phenomena. 53 illustrations. Preface. Index. 256pp. 5⅜ × 8½.
65533-4 Pa. $7.95

COMPLEX VARIABLES, Francis J. Flanigan. Unusual approach, delaying complex algebra till harmonic functions have been analyzed from real variable viewpoint. Includes problems with answers. 364pp. 5⅜ × 8½.   61388-7 Pa. $8.95

ATOMIC SPECTRA AND ATOMIC STRUCTURE, Gerhard Herzberg. One of best introductions; especially for specialist in other fields. Treatment is physical rather than mathematical. 80 illustrations. 257pp. 5⅜ × 8½.   60115-3 Pa. $6.95

APPLIED COMPLEX VARIABLES, John W. Dettman. Step-by-step coverage of fundamentals of analytic function theory—plus lucid exposition of five important applications: Potential Theory; Ordinary Differential Equations; Fourier Transforms; Laplace Transforms; Asymptotic Expansions. 66 figures. Exercises at chapter ends. 512pp. 5⅜ × 8½.   64670-X Pa. $11.95

ULTRASONIC ABSORPTION: An Introduction to the Theory of Sound Absorption and Dispersion in Gases, Liquids and Solids, A.B. Bhatia. Standard reference in the field provides a clear, systematically organized introductory review of fundamental concepts for advanced graduate students, research workers. Numerous diagrams. Bibliography. 440pp. 5⅜ × 8½.   64917-2 Pa. $11.95

UNBOUNDED LINEAR OPERATORS: Theory and Applications, Seymour Goldberg. Classic presents systematic treatment of the theory of unbounded linear operators in normed linear spaces with applications to differential equations. Bibliography. 199pp. 5⅜ × 8½.   64830-3 Pa. $7.95

LIGHT SCATTERING BY SMALL PARTICLES, H.C. van de Hulst. Comprehensive treatment including full range of useful approximation methods for researchers in chemistry, meteorology and astronomy. 44 illustrations. 470pp. 5⅜ × 8½.   64228-3 Pa. $11.95

CONFORMAL MAPPING ON RIEMANN SURFACES, Harvey Cohn. Lucid, insightful book presents ideal coverage of subject. 334 exercises make book perfect for self-study. 55 figures. 352pp. 5⅜ × 8¼.   64025-6 Pa. $9.95

OPTICKS, Sir Isaac Newton. Newton's own experiments with spectroscopy, colors, lenses, reflection, refraction, etc., in language the layman can follow. Foreword by Albert Einstein. 532pp. 5⅜ × 8½.   60205-2 Pa. $9.95

GENERALIZED INTEGRAL TRANSFORMATIONS, A.H. Zemanian. Graduate-level study of recent generalizations of the Laplace, Mellin, Hankel, K. Weierstrass, convolution and other simple transformations. Bibliography. 320pp. 5⅜ × 8½.   65375-7 Pa. $8.95

THE ELECTROMAGNETIC FIELD, Albert Shadowitz. Comprehensive undergraduate text covers basics of electric and magnetic fields, builds up to electromagnetic theory. Also related topics, including relativity. Over 900 problems. 768pp. 5⅜ × 8¼. 65660-8 Pa. $18.95

FOURIER SERIES, Georgi P. Tolstov. Translated by Richard A. Silverman. A valuable addition to the literature on the subject, moving clearly from subject to subject and theorem to theorem. 107 problems, answers. 336pp. 5⅜ × 8½. 63317-9 Pa. $8.95

THEORY OF ELECTROMAGNETIC WAVE PROPAGATION, Charles Herach Papas. Graduate-level study discusses the Maxwell field equations, radiation from wire antennas, the Doppler effect and more. xiii + 244pp. 5⅜ × 8½. 65678-0 Pa. $6.95

DISTRIBUTION THEORY AND TRANSFORM ANALYSIS: An Introduction to Generalized Functions, with Applications, A.H. Zemanian. Provides basics of distribution theory, describes generalized Fourier and Laplace transformations. Numerous problems. 384pp. 5⅜ × 8½. 65479-6 Pa. $9.95

THE PHYSICS OF WAVES, William C. Elmore and Mark A. Heald. Unique overview of classical wave theory. Acoustics, optics, electromagnetic radiation, more. Ideal as classroom text or for self-study. Problems. 477pp. 5⅜ × 8½. 64926-1 Pa. $12.95

CALCULUS OF VARIATIONS WITH APPLICATIONS, George M. Ewing. Applications-oriented introduction to variational theory develops insight and promotes understanding of specialized books, research papers. Suitable for advanced undergraduate/graduate students as primary, supplementary text. 352pp. 5⅜ × 8½. 64856-7 Pa. $8.95

A TREATISE ON ELECTRICITY AND MAGNETISM, James Clerk Maxwell. Important foundation work of modern physics. Brings to final form Maxwell's theory of electromagnetism and rigorously derives his general equations of field theory. 1,084pp. 5⅜ × 8½. 60636-8, 60637-6 Pa., Two-vol. set $21.90

AN INTRODUCTION TO THE CALCULUS OF VARIATIONS, Charles Fox. Graduate-level text covers variations of an integral, isoperimetrical problems, least action, special relativity, approximations, more. References. 279pp. 5⅜ × 8½. 65499-0 Pa. $7.95

HYDRODYNAMIC AND HYDROMAGNETIC STABILITY, S. Chandrasekhar. Lucid examination of the Rayleigh-Benard problem; clear coverage of the theory of instabilities causing convection. 704pp. 5⅜ × 8¼. 64071-X Pa. $14.95

CALCULUS OF VARIATIONS, Robert Weinstock. Basic introduction covering isoperimetric problems, theory of elasticity, quantum mechanics, electrostatics, etc. Exercises throughout. 326pp. 5⅜ × 8½. 63069-2 Pa. $8.95

DYNAMICS OF FLUIDS IN POROUS MEDIA, Jacob Bear. For advanced students of ground water hydrology, soil mechanics and physics, drainage and irrigation engineering and more. 335 illustrations. Exercises, with answers. 784pp. 6⅛ × 9¼. 65675-6 Pa. $19.95

NUMERICAL METHODS FOR SCIENTISTS AND ENGINEERS, Richard Hamming. Classic text stresses frequency approach in coverage of algorithms, polynomial approximation, Fourier approximation, exponential approximation, other topics. Revised and enlarged 2nd edition. 721pp. 5⅜ × 8½.
65241-6 Pa. $14.95

THEORETICAL SOLID STATE PHYSICS, Vol. I: Perfect Lattices in Equilibrium; Vol. II: Non-Equilibrium and Disorder, William Jones and Norman H. March. Monumental reference work covers fundamental theory of equilibrium properties of perfect crystalline solids, non-equilibrium properties, defects and disordered systems. Appendices. Problems. Preface. Diagrams. Index. Bibliography. Total of 1,301pp. 5⅜ × 8½. Two volumes.   Vol. I 65015-4 Pa. $14.95
Vol. II 65016-2 Pa. $14.95

OPTIMIZATION THEORY WITH APPLICATIONS, Donald A. Pierre. Broad-spectrum approach to important topic. Classical theory of minima and maxima, calculus of variations, simplex technique and linear programming, more. Many problems, examples. 640pp. 5⅜ × 8½.   65205-X Pa. $14.95

THE CONTINUUM: A Critical Examination of the Foundation of Analysis, Hermann Weyl. Classic of 20th-century foundational research deals with the conceptual problem posed by the continuum. 156pp. 5⅜ × 8½.   67982-9 Pa. $5.95

ESSAYS ON THE THEORY OF NUMBERS, Richard Dedekind. Two classic essays by great German mathematician: on the theory of irrational numbers; and on transfinite numbers and properties of natural numbers. 115pp. 5⅜ × 8½.
21010-3 Pa. $4.95

THE FUNCTIONS OF MATHEMATICAL PHYSICS, Harry Hochstadt. Comprehensive treatment of orthogonal polynomials, hypergeometric functions, Hill's equation, much more. Bibliography. Index. 322pp. 5⅜ × 8½.   65214-9 Pa. $9.95

NUMBER THEORY AND ITS HISTORY, Oystein Ore. Unusually clear, accessible introduction covers counting, properties of numbers, prime numbers, much more. Bibliography. 380pp. 5⅜ × 8½.   65620-9 Pa. $9.95

THE VARIATIONAL PRINCIPLES OF MECHANICS, Cornelius Lanczos. Graduate level coverage of calculus of variations, equations of motion, relativistic mechanics, more. First inexpensive paperbound edition of classic treatise. Index. Bibliography. 418pp. 5⅜ × 8½.   65067-7 Pa. $11.95

MATHEMATICAL TABLES AND FORMULAS, Robert D. Carmichael and Edwin R. Smith. Logarithms, sines, tangents, trig functions, powers, roots, reciprocals, exponential and hyperbolic functions, formulas and theorems. 269pp. 5⅜ × 8½.   60111-0 Pa. $6.95

THEORETICAL PHYSICS, Georg Joos, with Ira M. Freeman. Classic overview covers essential math, mechanics, electromagnetic theory, thermodynamics, quantum mechanics, nuclear physics, other topics. First paperback edition. xxiii + 885pp. 5⅜ × 8½.   65227-0 Pa. $19.95

HANDBOOK OF MATHEMATICAL FUNCTIONS WITH FORMULAS, GRAPHS, AND MATHEMATICAL TABLES, edited by Milton Abramowitz and Irene A. Stegun. Vast compendium: 29 sets of tables, some to as high as 20 places. 1,046pp. 8 × 10½. 61272-4 Pa. $24.95

MATHEMATICAL METHODS IN PHYSICS AND ENGINEERING, John W. Dettman. Algebraically based approach to vectors, mapping, diffraction, other topics in applied math. Also generalized functions, analytic function theory, more. Exercises. 448pp. 5⅜ × 8¼. 65649-7 Pa. **$9.95**

A SURVEY OF NUMERICAL MATHEMATICS, David M. Young and Robert Todd Gregory. Broad self-contained coverage of computer-oriented numerical algorithms for solving various types of mathematical problems in linear algebra, ordinary and partial, differential equations, much more. Exercises. Total of 1,248pp. 5⅜ × 8½. Two volumes. Vol. I 65691-8 Pa. $14.95
Vol. II 65692-6 Pa. $14.95

TENSOR ANALYSIS FOR PHYSICISTS, J.A. Schouten. Concise exposition of the mathematical basis of tensor analysis, integrated with well-chosen physical examples of the theory. Exercises. Index. Bibliography. 289pp. 5⅜ × 8½. 65582-2 Pa. $8.95

INTRODUCTION TO NUMERICAL ANALYSIS (2nd Edition), F.B. Hildebrand. Classic, fundamental treatment covers computation, approximation, interpolation, numerical differentiation and integration, other topics. 150 new problems. 669pp. 5⅜ × 8½. 65363-3 Pa. $15.95

INVESTIGATIONS ON THE THEORY OF THE BROWNIAN MOVEMENT, Albert Einstein. Five papers (1905–8) investigating dynamics of Brownian motion and evolving elementary theory. Notes by R. Fürth. 122pp. 5⅜ × 8½. 60304-0 Pa. $4.95

CATASTROPHE THEORY FOR SCIENTISTS AND ENGINEERS, Robert Gilmore. Advanced-level treatment describes mathematics of theory grounded in the work of Poincaré, R. Thom, other mathematicians. Also important applications to problems in mathematics, physics, chemistry and engineering. 1981 edition. References. 28 tables. 397 black-and-white illustrations. xvii + 666pp. 6⅛ × 9¼. 67539-4 Pa. $16.95

AN INTRODUCTION TO STATISTICAL THERMODYNAMICS, Terrell L. Hill. Excellent basic text offers wide-ranging coverage of quantum statistical mechanics, systems of interacting molecules, quantum statistics, more. 523pp. 5⅜ × 8½. 65242-4 Pa. $12.95

ELEMENTARY DIFFERENTIAL EQUATIONS, William Ted Martin and Eric Reissner. Exceptionally clear, comprehensive introduction at undergraduate level. Nature and origin of differential equations, differential equations of first, second and higher orders. Picard's Theorem, much more. Problems with solutions. 331pp. 5⅜ × 8½. 65024-3 Pa. $8.95

STATISTICAL PHYSICS, Gregory H. Wannier. Classic text combines thermodynamics, statistical mechanics and kinetic theory in one unified presentation of thermal physics. Problems with solutions. Bibliography. 532pp. 5⅜ × 8½. 65401-X Pa. $12.95

ORDINARY DIFFERENTIAL EQUATIONS, Morris Tenenbaum and Harry Pollard. Exhaustive survey of ordinary differential equations for undergraduates in mathematics, engineering, science. Thorough analysis of theorems. Diagrams. Bibliography. Index. 818pp. 5⅜ × 8½.   64940-7 Pa. $16.95

STATISTICAL MECHANICS: Principles and Applications, Terrell L. Hill. Standard text covers fundamentals of statistical mechanics, applications to fluctuation theory, imperfect gases, distribution functions, more. 448pp. 5⅜ × 8½.   65390-0 Pa. $11.95

ORDINARY DIFFERENTIAL EQUATIONS AND STABILITY THEORY: An Introduction, David A. Sánchez. Brief, modern treatment. Linear equation, stability theory for autonomous and nonautonomous systems, etc. 164pp. 5⅜ × 8¼.   63828-6 Pa. $5.95

THIRTY YEARS THAT SHOOK PHYSICS: The Story of Quantum Theory, George Gamow. Lucid, accessible introduction to influential theory of energy and matter. Careful explanations of Dirac's anti-particles, Bohr's model of the atom, much more. 12 plates. Numerous drawings. 240pp. 5⅜ × 8½.   24895-X Pa. $6.95

THEORY OF MATRICES, Sam Perlis. Outstanding text covering rank, non-singularity and inverses in connection with the development of canonical matrices under the relation of equivalence, and without the intervention of determinants. Includes exercises. 237pp. 5⅜ × 8½.   66810-X Pa. $7.95

GREAT EXPERIMENTS IN PHYSICS: Firsthand Accounts from Galileo to Einstein, edited by Morris H. Shamos. 25 crucial discoveries: Newton's laws of motion, Chadwick's study of the neutron, Hertz on electromagnetic waves, more. Original accounts clearly annotated. 370pp. 5⅜ × 8½.   25346-5 Pa. $10.95

INTRODUCTION TO PARTIAL DIFFERENTIAL EQUATIONS WITH AP-PLICATIONS, E.C. Zachmanoglou and Dale W. Thoe. Essentials of partial differential equations applied to common problems in engineering and the physical sciences. Problems and answers. 416pp. 5⅜ × 8½.   65251-3 Pa. $10.95

BURNHAM'S CELESTIAL HANDBOOK, Robert Burnham, Jr. Thorough guide to the stars beyond our solar system. Exhaustive treatment. Alphabetical by constellation: Andromeda to Cetus in Vol. 1; Chamaeleon to Orion in Vol. 2; and Pavo to Vulpecula in Vol. 3. Hundreds of illustrations. Index in Vol. 3. 2,000pp. 6⅛ × 9¼.   23567-X, 23568-8, 23673-0 Pa., Three-vol. set $41.85

CHEMICAL MAGIC, Leonard A. Ford. Second Edition, Revised by E. Winston Grundmeier. Over 100 unusual stunts demonstrating cold fire, dust explosions, much more. Text explains scientific principles and stresses safety precautions. 128pp. 5⅜ × 8½.   67628-5 Pa. $5.95

AMATEUR ASTRONOMER'S HANDBOOK, J.B. Sidgwick. Timeless, comprehensive coverage of telescopes, mirrors, lenses, mountings, telescope drives, micrometers, spectroscopes, more. 189 illustrations. 576pp. 5⅜ × 8¼. (Available in U.S. only)   24034-7 Pa. $9.95

SPECIAL FUNCTIONS, N.N. Lebedev. Translated by Richard Silverman. Famous Russian work treating more important special functions, with applications to specific problems of physics and engineering. 38 figures. 308pp. 5⅜ × 8½.
60624-4 Pa. $8.95

OBSERVATIONAL ASTRONOMY FOR AMATEURS, J.B. Sidgwick. Mine of useful data for observation of sun, moon, planets, asteroids, aurorae, meteors, comets, variables, binaries, etc. 39 illustrations. 384pp. 5⅜ × 8¼. (Available in U.S. only)
24033-9 Pa. $8.95

INTEGRAL EQUATIONS, F.G. Tricomi. Authoritative, well-written treatment of extremely useful mathematical tool with wide applications. Volterra Equations, Fredholm Equations, much more. Advanced undergraduate to graduate level. Exercises. Bibliography. 238pp. 5⅜ × 8½.
64828-1 Pa. $7.95

POPULAR LECTURES ON MATHEMATICAL LOGIC, Hao Wang. Noted logician's lucid treatment of historical developments, set theory, model theory, recursion theory and constructivism, proof theory, more. 3 appendixes. Bibliography. 1981 edition. ix + 283pp. 5⅜ × 8½.
67632-3 Pa. $8.95

MODERN NONLINEAR EQUATIONS, Thomas L. Saaty. Emphasizes practical solution of problems; covers seven types of equations. ". . . a welcome contribution to the existing literature. . . ."—*Math Reviews.* 490pp. 5⅜ × 8½. 64232-1 Pa. $11.95

FUNDAMENTALS OF ASTRODYNAMICS, Roger Bate et al. Modern approach developed by U.S. Air Force Academy. Designed as a first course. Problems, exercises. Numerous illustrations. 455pp. 5⅜ × 8½. 60061-0 Pa. $9.95

INTRODUCTION TO LINEAR ALGEBRA AND DIFFERENTIAL EQUATIONS, John W. Dettman. Excellent text covers complex numbers, determinants, orthonormal bases, Laplace transforms, much more. Exercises with solutions. Undergraduate level. 416pp. 5⅜ × 8½. 65191-6 Pa. $10.95

INCOMPRESSIBLE AERODYNAMICS, edited by Bryan Thwaites. Covers theoretical and experimental treatment of the uniform flow of air and viscous fluids past two-dimensional aerofoils and three-dimensional wings; many other topics. 654pp. 5⅜ × 8½. 65465-6 Pa. $16.95

INTRODUCTION TO DIFFERENCE EQUATIONS, Samuel Goldberg. Exceptionally clear exposition of important discipline with applications to sociology, psychology, economics. Many illustrative examples; over 250 problems. 260pp. 5⅜ × 8½. 65084-7 Pa. $7.95

LAMINAR BOUNDARY LAYERS, edited by L. Rosenhead. Engineering classic covers steady boundary layers in two- and three-dimensional flow, unsteady boundary layers, stability, observational techniques, much more. 708pp. 5⅜ × 8½.
65646-2 Pa. $18.95

LECTURES ON CLASSICAL DIFFERENTIAL GEOMETRY, Second Edition, Dirk J. Struik. Excellent brief introduction covers curves, theory of surfaces, fundamental equations, geometry on a surface, conformal mapping, other topics. Problems. 240pp. 5⅜ × 8½. 65609-8 Pa. $8.95

CATALOG OF DOVER BOOKS

ROTARY-WING AERODYNAMICS, W.Z. Stepniewski. Clear, concise text covers aerodynamic phenomena of the rotor and offers guidelines for helicopter performance evaluation. Originally prepared for NASA. 537 figures. 640pp. 6⅛ × 9¼.
64647-5 Pa. $15.95

DIFFERENTIAL GEOMETRY, Heinrich W. Guggenheimer. Local differential geometry as an application of advanced calculus and linear algebra. Curvature, transformation groups, surfaces, more. Exercises. 62 figures. 378pp. 5⅜ × 8½.
63433-7 Pa. $8.95

INTRODUCTION TO SPACE DYNAMICS, William Tyrrell Thomson. Comprehensive, classic introduction to space-flight engineering for advanced undergraduate and graduate students. Includes vector algebra, kinematics, transformation of coordinates. Bibliography. Index. 352pp. 5⅜ × 8½. 65113-4 Pa. $8.95

A SURVEY OF MINIMAL SURFACES, Robert Osserman. Up-to-date, in-depth discussion of the field for advanced students. Corrected and enlarged edition covers new developments. Includes numerous problems. 192pp. 5⅜ × 8½.
64998-9 Pa. $8.95

ANALYTICAL MECHANICS OF GEARS, Earle Buckingham. Indispensable reference for modern gear manufacture covers conjugate gear-tooth action, gear-tooth profiles of various gears, many other topics. 263 figures. 102 tables. 546pp. 5⅜ × 8½. 65712-4 Pa. $14.95

SET THEORY AND LOGIC, Robert R. Stoll. Lucid introduction to unified theory of mathematical concepts. Set theory and logic seen as tools for conceptual understanding of real number system. 496pp. 5⅜ × 8¼. 63829-4 Pa. $12.95

A HISTORY OF MECHANICS, René Dugas. Monumental study of mechanical principles from antiquity to quantum mechanics. Contributions of ancient Greeks, Galileo, Leonardo, Kepler, Lagrange, many others. 671pp. 5⅜ × 8½.
65632-2 Pa. $14.95

FAMOUS PROBLEMS OF GEOMETRY AND HOW TO SOLVE THEM, Benjamin Bold. Squaring the circle, trisecting the angle, duplicating the cube: learn their history, why they are impossible to solve, then solve them yourself. 128pp. 5⅜ × 8½. 24297-8 Pa. $4.95

MECHANICAL VIBRATIONS, J.P. Den Hartog. Classic textbook offers lucid explanations and illustrative models, applying theories of vibrations to a variety of practical industrial engineering problems. Numerous figures. 233 problems, solutions. Appendix. Index. Preface. 436pp. 5⅜ × 8½. 64785-4 Pa. $10.95

CURVATURE AND HOMOLOGY, Samuel I. Goldberg. Thorough treatment of specialized branch of differential geometry. Covers Riemannian manifolds, topology of differentiable manifolds, compact Lie groups, other topics. Exercises. 315pp. 5⅜ × 8½. 64314-X Pa. $9.95

HISTORY OF STRENGTH OF MATERIALS, Stephen P. Timoshenko. Excellent historical survey of the strength of materials with many references to the theories of elasticity and structure. 245 figures. 452pp. 5⅜ × 8½. 61187-6 Pa. $11.95

GEOMETRY OF COMPLEX NUMBERS, Hans Schwerdtfeger. Illuminating, widely praised book on analytic geometry of circles, the Moebius transformation, and two-dimensional non-Euclidean geometries. 200pp. 5⅜ × 8¼.
63830-8 Pa. $8.95

MECHANICS, J.P. Den Hartog. A classic introductory text or refresher. Hundreds of applications and design problems illuminate fundamentals of trusses, loaded beams and cables, etc. 334 answered problems. 462pp. 5⅜ × 8½. 60754-2 Pa. $9.95

TOPOLOGY, John G. Hocking and Gail S. Young. Superb one-year course in classical topology. Topological spaces and functions, point-set topology, much more. Examples and problems. Bibliography. Index. 384pp. 5⅜ × 8¼.
65676-4 Pa. $9.95

STRENGTH OF MATERIALS, J.P. Den Hartog. Full, clear treatment of basic material (tension, torsion, bending, etc.) plus advanced material on engineering methods, applications. 350 answered problems. 323pp. 5⅜ × 8½. 60755-0 Pa. $8.95

ELEMENTARY CONCEPTS OF TOPOLOGY, Paul Alexandroff. Elegant, intuitive approach to topology from set-theoretic topology to Betti groups; how concepts of topology are useful in math and physics. 25 figures. 57pp. 5⅜ × 8½.
60747-X Pa. $3.50

ADVANCED STRENGTH OF MATERIALS, J.P. Den Hartog. Superbly written advanced text covers torsion, rotating disks, membrane stresses in shells, much more. Many problems and answers. 388pp. 5⅜ × 8½. 65407-9 Pa. $9.95

COMPUTABILITY AND UNSOLVABILITY, Martin Davis. Classic graduate-level introduction to theory of computability, usually referred to as theory of recurrent functions. New preface and appendix. 288pp. 5⅜ × 8½. 61471-9 Pa. $7.95

GENERAL CHEMISTRY, Linus Pauling. Revised 3rd edition of classic first-year text by Nobel laureate. Atomic and molecular structure, quantum mechanics, statistical mechanics, thermodynamics correlated with descriptive chemistry. Problems. 992pp. 5⅜ × 8½. 65622-5 Pa. $19.95

AN INTRODUCTION TO MATRICES, SETS AND GROUPS FOR SCIENCE STUDENTS, G. Stephenson. Concise, readable text introduces sets, groups, and most importantly, matrices to undergraduate students of physics, chemistry, and engineering. Problems. 164pp. 5⅜ × 8½. 65077-4 Pa. $6.95

THE HISTORICAL BACKGROUND OF CHEMISTRY, Henry M. Leicester. Evolution of ideas, not individual biography. Concentrates on formulation of a coherent set of chemical laws. 260pp. 5⅜ × 8½. 61053-5 Pa. $6.95

THE PHILOSOPHY OF MATHEMATICS: An Introductory Essay, Stephan Körner. Surveys the views of Plato, Aristotle, Leibniz & Kant concerning propositions and theories of applied and pure mathematics. Introduction. Two appendices. Index. 198pp. 5⅜ × 8½. 25048-2 Pa. $7.95

THE DEVELOPMENT OF MODERN CHEMISTRY, Aaron J. Ihde. Authoritative history of chemistry from ancient Greek theory to 20th-century innovation. Covers major chemists and their discoveries. 209 illustrations. 14 tables. Bibliographies. Indices. Appendices. 851pp. 5⅜ × 8½. 64235-6 Pa. $18.95

DE RE METALLICA, Georgius Agricola. The famous Hoover translation of greatest treatise on technological chemistry, engineering, geology, mining of early modern times (1556). All 289 original woodcuts. 638pp. 6¾ × 11.

60006-8 Pa. $18.95

SOME THEORY OF SAMPLING, William Edwards Deming. Analysis of the problems, theory and design of sampling techniques for social scientists, industrial managers and others who find statistics increasingly important in their work. 61 tables. 90 figures. xvii + 602pp. 5⅜ × 8½.

64684-X Pa. $15.95

THE VARIOUS AND INGENIOUS MACHINES OF AGOSTINO RAMELLI: A Classic Sixteenth-Century Illustrated Treatise on Technology, Agostino Ramelli. One of the most widely known and copied works on machinery in the 16th century. 194 detailed plates of water pumps, grain mills, cranes, more. 608pp. 9 × 12.

28180-9 Pa. $24.95

LINEAR PROGRAMMING AND ECONOMIC ANALYSIS, Robert Dorfman, Paul A. Samuelson and Robert M. Solow. First comprehensive treatment of linear programming in standard economic analysis. Game theory, modern welfare economics, Leontief input-output, more. 525pp. 5⅜ × 8½.

65491-5 Pa. $14.95

ELEMENTARY DECISION THEORY, Herman Chernoff and Lincoln E. Moses. Clear introduction to statistics and statistical theory covers data processing, probability and random variables, testing hypotheses, much more. Exercises. 364pp. 5⅜ × 8½.

65218-1 Pa. $9.95

THE COMPLEAT STRATEGYST: Being a Primer on the Theory of Games of Strategy, J.D. Williams. Highly entertaining classic describes, with many illustrated examples, how to select best strategies in conflict situations. Prefaces. Appendices. 268pp. 5⅜ × 8½.

25101-2 Pa. $7.95

MATHEMATICAL METHODS OF OPERATIONS RESEARCH, Thomas L. Saaty. Classic graduate-level text covers historical background, classical methods of forming models, optimization, game theory, probability, queueing theory, much more. Exercises. Bibliography. 448pp. 5⅜ × 8¼.

65703-5 Pa. $12.95

CONSTRUCTIONS AND COMBINATORIAL PROBLEMS IN DESIGN OF EXPERIMENTS, Damaraju Raghavarao. In-depth reference work examines orthogonal Latin squares, incomplete block designs, tactical configuration, partial geometry, much more. Abundant explanations, examples. 416pp. 5⅜ × 8¼.

65685-3 Pa. $10.95

THE ABSOLUTE DIFFERENTIAL CALCULUS (CALCULUS OF TENSORS), Tullio Levi-Civita. Great 20th-century mathematician's classic work on material necessary for mathematical grasp of theory of relativity. 452pp. 5⅜ × 8½.

63401-9 Pa. $9.95

VECTOR AND TENSOR ANALYSIS WITH APPLICATIONS, A.I. Borisenko and I.E. Tarapov. Concise introduction. Worked-out problems, solutions, exercises. 257pp. 5⅜ × 8¼.

63833-2 Pa. $7.95

THE FOUR-COLOR PROBLEM: Assaults and Conquest, Thomas L. Saaty and Paul G. Kainen. Engrossing, comprehensive account of the century-old combinatorial topological problem, its history and solution. Bibliographies. Index. 110 figures. 228pp. 5⅜ × 8½. 65092-8 Pa. $6.95

CATALYSIS IN CHEMISTRY AND ENZYMOLOGY, William P. Jencks. Exceptionally clear coverage of mechanisms for catalysis, forces in aqueous solution, carbonyl- and acyl-group reactions, practical kinetics, more. 864pp. 5⅜ × 8½. 65460-5 Pa. $19.95

PROBABILITY: An Introduction, Samuel Goldberg. Excellent basic text covers set theory, probability theory for finite sample spaces, binomial theorem, much more. 360 problems. Bibliographies. 322pp. 5⅜ × 8½. 65252-1 Pa. $8.95

LIGHTNING, Martin A. Uman. Revised, updated edition of classic work on the physics of lightning. Phenomena, terminology, measurement, photography, spectroscopy, thunder, more. Reviews recent research. Bibliography. Indices. 320pp. 5⅜ × 8¼. 64575-4 Pa. $8.95

PROBABILITY THEORY: A Concise Course, Y.A. Rozanov. Highly readable, self-contained introduction covers combination of events, dependent events, Bernoulli trials, etc. Translation by Richard Silverman. 148pp. 5⅜ × 8¼.
63544-9 Pa. $5.95

AN INTRODUCTION TO HAMILTONIAN OPTICS, H. A. Buchdahl. Detailed account of the Hamiltonian treatment of aberration theory in geometrical optics. Many classes of optical systems defined in terms of the symmetries they possess. Problems with detailed solutions. 1970 edition. xv + 360pp. 5⅜ × 8½.
67597-1 Pa. $10.95

STATISTICS MANUAL, Edwin L. Crow, et al. Comprehensive, practical collection of classical and modern methods prepared by U.S. Naval Ordnance Test Station. Stress on use. Basics of statistics assumed. 288pp. 5⅜ × 8½.
60599-X Pa. $6.95

DICTIONARY/OUTLINE OF BASIC STATISTICS, John E. Freund and Frank J. Williams. A clear concise dictionary of over 1,000 statistical terms and an outline of statistical formulas covering probability, nonparametric tests, much more. 208pp. 5⅜ × 8½. 66796-0 Pa. $6.95

STATISTICAL METHOD FROM THE VIEWPOINT OF QUALITY CONTROL, Walter A. Shewhart. Important text explains regulation of variables, uses of statistical control to achieve quality control in industry, agriculture, other areas. 192pp. 5⅜ × 8½. 65232-7 Pa. $7.95

THE INTERPRETATION OF GEOLOGICAL PHASE DIAGRAMS, Ernest G. Ehlers. Clear, concise text emphasizes diagrams of systems under fluid or containing pressure; also coverage of complex binary systems, hydrothermal melting, more. 288pp. 6½ × 9¼. 65389-7 Pa. $10.95

STATISTICAL ADJUSTMENT OF DATA, W. Edwards Deming. Introduction to basic concepts of statistics, curve fitting, least squares solution, conditions without parameter, conditions containing parameters. 26 exercises worked out. 271pp. 5⅜ × 8½. 64685-8 Pa. $8.95

TENSOR CALCULUS, J.L. Synge and A. Schild. Widely used introductory text covers spaces and tensors, basic operations in Riemannian space, non-Riemannian spaces, etc. 324pp. 5⅜ × 8¼. 63612-7 Pa. $8.95

A CONCISE HISTORY OF MATHEMATICS, Dirk J. Struik. The best brief history of mathematics. Stresses origins and covers every major figure from ancient Near East to 19th century. 41 illustrations. 195pp. 5⅜ × 8½. 60255-9 Pa. $7.95

A SHORT ACCOUNT OF THE HISTORY OF MATHEMATICS, W.W. Rouse Ball. One of clearest, most authoritative surveys from the Egyptians and Phoenicians through 19th-century figures such as Grassman, Galois, Riemann. Fourth edition. 522pp. 5⅜ × 8½. 20630-0 Pa. $10.95

HISTORY OF MATHEMATICS, David E. Smith. Nontechnical survey from ancient Greece and Orient to late 19th century; evolution of arithmetic, geometry, trigonometry, calculating devices, algebra, the calculus. 362 illustrations. 1,355pp. 5⅜ × 8½. 20429-4, 20430-8 Pa., Two-vol. set $23.90

THE GEOMETRY OF RENÉ DESCARTES, René Descartes. The great work founded analytical geometry. Original French text, Descartes' own diagrams, together with definitive Smith-Latham translation. 244pp. 5⅜ × 8½.
60068-8 Pa. $7.95

THE ORIGINS OF THE INFINITESIMAL CALCULUS, Margaret E. Baron. Only fully detailed and documented account of crucial discipline: origins; development by Galileo, Kepler, Cavalieri; contributions of Newton, Leibniz, more. 304pp. 5⅜ × 8½. (Available in U.S. and Canada only) 65371-4 Pa. $9.95

THE HISTORY OF THE CALCULUS AND ITS CONCEPTUAL DEVELOP-MENT, Carl B. Boyer. Origins in antiquity, medieval contributions, work of Newton, Leibniz, rigorous formulation. Treatment is verbal. 346pp. 5⅜ × 8½.
60509-4 Pa. $8.95

THE THIRTEEN BOOKS OF EUCLID'S ELEMENTS, translated with introduction and commentary by Sir Thomas L. Heath. Definitive edition. Textual and linguistic notes, mathematical analysis. 2,500 years of critical commentary. Not abridged. 1,414pp. 5⅜ × 8½. 60088-2, 60089-0, 60090-4 Pa., Three-vol. set $29.85

GAMES AND DECISIONS: Introduction and Critical Survey, R. Duncan Luce and Howard Raiffa. Superb nontechnical introduction to game theory, primarily applied to social sciences. Utility theory, zero-sum games, n-person games, decision-making, much more. Bibliography. 509pp. 5⅜ × 8½. 65943-7 Pa. $12.95

THE HISTORICAL ROOTS OF ELEMENTARY MATHEMATICS, Lucas N.H. Bunt, Phillip S. Jones, and Jack D. Bedient. Fundamental underpinnings of modern arithmetic, algebra, geometry and number systems derived from ancient civilizations. 320pp. 5⅜ × 8½. 25563-8 Pa. $8.95

CALCULUS REFRESHER FOR TECHNICAL PEOPLE, A. Albert Klaf. Covers important aspects of integral and differential calculus via 756 questions. 566 problems, most answered. 431pp. 5⅜ × 8½. 20370-0 Pa. $8.95

CATALOG OF DOVER BOOKS

CHALLENGING MATHEMATICAL PROBLEMS WITH ELEMENTARY SOLUTIONS, A.M. Yaglom and I.M. Yaglom. Over 170 challenging problems on probability theory, combinatorial analysis, points and lines, topology, convex polygons, many other topics. Solutions. Total of 445pp. 5⅜ × 8½. Two-vol. set.

Vol. I 65536-9 Pa. $7.95
Vol. II 65537-7 Pa. $6.95

FIFTY CHALLENGING PROBLEMS IN PROBABILITY WITH SOLUTIONS, Frederick Mosteller. Remarkable puzzlers, graded in difficulty, illustrate elementary and advanced aspects of probability. Detailed solutions. 88pp. 5⅜ × 8½.

65355-2 Pa. $4.95

EXPERIMENTS IN TOPOLOGY, Stephen Barr. Classic, lively explanation of one of the byways of mathematics. Klein bottles, Moebius strips, projective planes, map coloring, problem of the Koenigsberg bridges, much more, described with clarity and wit. 43 figures. 210pp. 5⅜ × 8½.

25933-1 Pa. $5.95

RELATIVITY IN ILLUSTRATIONS, Jacob T. Schwartz. Clear nontechnical treatment makes relativity more accessible than ever before. Over 60 drawings illustrate concepts more clearly than text alone. Only high school geometry needed. Bibliography. 128pp. 6⅛ × 9¼.

25965-X Pa. $6.95

AN INTRODUCTION TO ORDINARY DIFFERENTIAL EQUATIONS, Earl A. Coddington. A thorough and systematic first course in elementary differential equations for undergraduates in mathematics and science, with many exercises and problems (with answers). Index. 304pp. 5⅜ × 8½.

65942-9 Pa. $8.95

FOURIER SERIES AND ORTHOGONAL FUNCTIONS, Harry F. Davis. An incisive text combining theory and practical example to introduce Fourier series, orthogonal functions and applications of the Fourier method to boundary-value problems. 570 exercises. Answers and notes. 416pp. 5⅜ × 8½.

65973-9 Pa. $9.95

THE THEORY OF BRANCHING PROCESSES, Theodore E. Harris. First systematic, comprehensive treatment of branching (i.e. multiplicative) processes and their applications. Galton-Watson model, Markov branching processes, electron-photon cascade, many other topics. Rigorous proofs. Bibliography. 240pp. 5⅜ × 8½.

65952-6 Pa. $6.95

AN INTRODUCTION TO ALGEBRAIC STRUCTURES, Joseph Landin. Superb self-contained text covers "abstract algebra": sets and numbers, theory of groups, theory of rings, much more. Numerous well-chosen examples, exercises. 247pp. 5⅜ × 8½.

65940-2 Pa. $7.95

Prices subject to change without notice.
Available at your book dealer or write for free Mathematics and Science Catalog to Dept. GI, Dover Publications, Inc., 31 East 2nd St., Mineola, N.Y. 11501. Dover publishes more than 175 books each year on science, elementary and advanced mathematics, biology, music, art, literature, history, social sciences and other areas.